MW01518514

Remediation of Hazardous Waste in the Subsurface

250101
#180.95

ACS SYMPOSIUM SERIES **940**

Remediation of Hazardous Waste in the Subsurface

Bridging Flask and Field

Clayton J. Clark II, Editor
University of Florida

Angela Stephenson Lindner, Editor
University of Florida

Sponsored by the
ACS Division of Environmental Chemistry, Inc.

American Chemical Society, Washington, DC

Remediation of hazardous waste in the subsurface
c2006.

ISBN-13: 978–0–8412–3969–2
ISBN-10: 0–8412–3969–X

The paper used in this publication meets the minimum requirements of American National Standard for Information Sciences—Permanence of Paper for Printed Library Materials, ANSI Z39.48–1984.

Distributed by Oxford University Press

PRINTED IN THE UNITED STATES OF AMERICA

Foreword

The ACS Symposium Series was first published in 1974 to provide a mechanism for publishing symposia quickly in book form. The purpose of the series is to publish timely, comprehensive books developed from ACS sponsored symposia based on current scientific research. Occasionally, books are developed from symposia sponsored by other organizations when the topic is of keen interest to the chemistry audience.

Before agreeing to publish a book, the proposed table of contents is reviewed for appropriate and comprehensive coverage and for interest to the audience. Some papers may be excluded to better focus the book; others may be added to provide comprehensiveness. When appropriate, overview or introductory chapters are added. Drafts of chapters are peer-reviewed prior to final acceptance or rejection, and manuscripts are prepared in camera-ready format.

As a rule, only original research papers and original review papers are included in the volumes. Verbatim reproductions of previously published papers are not accepted.

ACS Books Department

Contents

Introduction

Physicochemical Methods of Subsurface Remediation

Biological Methods of Subsurface Remediation

Modeling in Physical, Chemical, and Biological Methods

Indexes

Preface

This book includes the work of researchers who have linked laboratory- and field-scales in efforts to find creative, cost-effective methods for successful remediation of contaminated soil and groundwater. The chapters in this book focus on remediation of the subsurface through using physicochemical or biological methods and modeling of the subsurface with both the field and laboratory scales in mind. Thus, this book is divided into three major sections: physicochemical remediation approaches, biological remediation approaches, and modeling approaches. Chapters in this book include descriptions of innovative methods that have been or are being developed for field-scale implementation. It was not the goal in the construction of this book to provide a comprehensive discussion of the most recent technologies being considered for restoring subsurface sites that are contaminated with hazardous wastes, but rather to begin a dialogue of how the flask and the field can be combined for improved diagnosis of a remediation method's effectiveness, efficiency, and economy. Overall, the book serves two purposes: (1) to provide an updated review of the most current subsurface treatment methodologies and (2) to present tangible methods that ultimately will enable the environmental practitioner to assess field treatment effectiveness by using laboratory-based testing.

Due to the comprehensive breadth of this book, chemists and engineers who are involved in remediation of hazardous waste and research applied in this area will be potential end-users of this book. Specific fields of employment of this audience include environmental consulting, state and federal governments, chemical industry, and academia. Others who are likely to benefit from this book include students of chemistry and engineering and chemists and engineers who need to become familiarized with subsurface remediation methods at some point during their careers.

The editors thank all of the authors and their respective staffs for their assistance in compiling this project. The varying expertise displayed in this book is a testament to their hard work, diligence, and cooperation. The editors also thank their Father in Heaven through their Lord and Savior Jesus Christ for His direction and support throughout the completion of this project.

Clayton J. Clark II
Department of Civil and Coastal Engineering
University of Florida
P.O. Box 116580
Gainesville, FL 32611–2083

Angela Stephenson Lindner
Department of Environmental Engineering Sciences
University of Florida
P.O. Box 116450
Gainesville, FL 32611–6450

Acknowledgments

The editors thank all of the authors and their respective staffs for their assistance in compiling this project. The varying expertise displayed in this book is a testament to their hard work, diligence, and cooperation. The editors also thank their Father in Heaven through their Lord and Savior Jesus Christ for His direction and support throughout the completion of this project.

Dr. Clark also thanks his wife, Kimberline, and their sons, Clayton Joshua III and Chayce Othniel, for their love and prayers. He also thanks his parents, Drs. Clayton J. and A. T. Clark, and his entire family, by blood and by Spirit, for their love and support. He also thanks all colleagues and friends and expresses to them that whatever help, no matter how small, was truly appreciated.

Dr. Lindner deeply acknowledges the unending physical and emotional support of her husband, James. To her mother, Minerva Stephenson, and brother, Ralph Stephenson Jr., she gives her thanks for their keeping her on course in living a faithful and courageous life in the shadow of adversity. To Dad and Katy, you are always in my heart.

Remediation of Hazardous Waste
in the Subsurface

Introduction

Chapter 1

Remediation of Hazardous Waste in the Subsurface: Bridging Flask and Field: Introduction

Clayton J. Clark II[1] and Angela Stephenson Lindner[2]

Departments of [1]Civil Engineering and [2]Environmental Engineering
Sciences, University of Florida, Gainesville, FL 32611

Rationale of This Book

Despite over 25 years of development of subsurface remediation technologies and an increasing number of resources dedicated to disseminate information about what has been learned in the process, cleanup of subsurface environments remains as one of the most difficult challenges for the environmental community today. A recent report by the U.S. Environmental Protection Agency has estimated that between 235,000 and 355,000 sites contaminated by hazardous waste in the United States will require between 30-35 years and from $170-250 million to complete most of the work needed (*1*), and a significant portion of these sites involve subsurface contamination.

According to the Resource Conservation and Recovery Act (RCRA), a hazardous waste is defined as a substance possessing any one of the following characteristics: *ignitability, reactivity, corrosivity,* and *toxicity.* Major classes of hazardous waste contaminants include metals, volatile organic compounds (VOCs), including benzene, toluene, ethylbenzene and xylene (BTEX), halogenated compounds, polyaromatic and hydrocarbons (PAHs), semivolatile organic compounds (SVOCs), including halogenated and non-halogenated compounds, phenols, pesticides, and polychlorinated biphenyls (PCBs). VOCs and metals tend to be the most prevalent contaminants in hazardous waste sites;

however, SVOCs still account for a significant portion of waste. Management of these wastes at contaminated sites in the United States is overseen by the seven major cleanup programs: Superfund (National Priority List, NPL), RCRA Corrective Action, Underground Storage Tanks (UST), Department of Defense (DOD), Department of Energy (DOE), Other (Civilian) Federal Agencies, and States and Private Parties. Table I provides a summary of the number of sites and the percent media contamination present in four of these programs (1). Data of this type are not available for the other programs.

Most of the cleanup cost of these sites is borne by the owners of the properties (in the case of private and public entities) or by those potentially responsible for the contamination (1). The Superfund Program has earned a reputation of being a leader in use of innovative, efficient, and economical technologies for site cleanup (1). Forty-two percent of all treatments chosen in the Superfund Program for remediation of soils, sediments, sludge and solid-matrix wastes have involved *in situ* technologies, such as soil vapor extraction

Table I. Summary of Sites and Media Contamination Overseen by Major United States Cleanup Programs (1)

Program	Number of Sites[a]	Percentage of Sites		
		Groundwater	Soil	Sediment
Superfund (NPL)	736	83	78	32
RCRA	3,800	82	61	6
DOD	6,400	63	77	18
DOE	5,000	72	72	72
UST	125,000	Not available	Not available	Not available
Civilian Federal Agencies	>3000	Not available	Not available	Not available
States and Private Parties	150,000	Not available	Not available	Not available

[a]These values vary with time and include only sites requiring cleanup and not where cleanup is ongoing or complete.

and bioremediation, whereas the remainder of treatments have been *ex situ* in nature, including solidification/stabilization and incineration. The most commonly selected groundwater remedies at NPL sites have been pump and treat methods (67%), monitored natural attenuation (19%), and *in situ* treatment (13%) (*1*). Although the seven major programs have had tremendous success in reducing immediate threats of contaminated sites, their ability to effect restoration to drinking water quality, for example, has been severely hampered by site-to-site variability in hydrogeologic characteristics and contaminant properties, and inadequate remediation system design and implementation (*2, 3*).

The inevitable concern of ineffective methods of restoring hazardous waste sites is their risk to public health. Assessing the NPL sites alone between 1992 and 1996, the Agency for Toxic Substance Disease Registry estimated that an average of 46% of these sites were considered as "urgent hazards" or "health hazards," the two highest categories of the total of six categories used in health assessments (*4*). As a result, there is urgent need for not only effective methods but also more economical means to facilitate faster remediation.

Restriction of the remediation process and accompanying performance testing before, during, and after treatment to the field scale alone is accompanied by inherent benefits and weaknesses. While field scale testing is most realistic, problems may arise from the intractability of the large-scale experiments and the longer duration requirements (*5*). Laboratory-based studies, such as microcosm experiments, are tractable, can be replicated for statistical analyses, and can be maintained under constant conditions. Integration of these two levels, accompanied by modeling, provides the most powerful approach for site cleanup (*5*). However, in the recent past due to time and cost considerations, a lack of value placed in laboratory and modeling studies, and lack of sophisticated, rapid diagnostic tools, there has been reluctance to bridge field and flask. Today, rapid advances in laboratory and modeling methods are occurring. Given the increasing number of hazardous wastes sites and their complexity, there is great need to develop paths that link these methods to the field to facilitate restoration of contaminated sites.

Discussion

The chapters in this book focus on remediation of the subsurface using physicochemical or biological methods and modeling of the subsurface with both the field and laboratory scales in mind. Thus, this book is divided into three major sections: Physicochemical Remediation Approaches, Biological Remediation Approaches, and Modeling Approaches. Chapters in this book include descriptions of innovative methods that have been or are being developed for field-scale implementation. It was not the goal in the construction of this book to provide a comprehensive discussion of the most recent technologies being considered for restoring subsurface sites contaminated with hazardous wastes but, rather, to begin a dialogue of how the flask and the field can be combined for improved diagnosis of a remediation method's effectiveness, efficiency, and economy. The contents of this book are discussed below.

Physicochemical Remediation Approaches

Crapse et al. report findings of studies of natural attenuation of beryllium at a field site impacted by coal plant operations. By conducting both laboratory and field studies, they observed a direct connection between elevated pH levels and higher soil concentrations of beryllium (and higher percentage in the available fraction), thus increasing our understanding of the partitioning mechanisms of this metal.

While cosolvent flushing has been used effectively for various chlorinated contaminants in the subsurface, Chen and Clark report the effectiveness of cosolvent methanol in removing toxaphene, a once prevalently used pesticide, from column soils. They present agreement of their experimental and modeling results in the log-linear relationships between the toxaphene sorption constant K_p and methanol fraction f_c, the reverse first-order rate constant k_2 and f_c, and k_2 and K_p. These results show tremendous potential for removal of persistent toxaphene from contaminated sites.

De Las Casas et al. report the effectiveness of Fenton's reagents in regeneration of spent granulated activated carbon (GAC) used for removal of a variety of chlorinated hydrocarbons both in the laboratory and in the field. Eighty-nine percent of adsorbed methylene chloride was removed from the laboratory system after regeneration, whereas 95% of trichloroethylene was shown to be removed at the field site. The authors report that, while the method has not proven to be cost-effective in relation to other GAC regeneration methods, alterations in design and operation may potentially lower process costs.

Gerstl et al. report the novel use of short- and long-chain organoclays with organophilic surfaces for sorption of a variety of pesticides and polyaromatic hydrocarbons from the aqueous phase. The short-chain organoclays were most effective in removing compounds with low-to-medium hydrophobicities, whereas the long-chain organoclays were most effective in removing strongly hydrophobic compounds. The authors also report the performance of a bifunctional organoclay, able to both adsorb and catalyze the hydrolysis of organophosphate pesticides with significant implications for destruction of these contaminants in the field.

Biological Remediation Approaches

By careful monitoring of carbon dioxide, along with soil temperature and oxygen supply, Berry et al. report success in removing petroleum from radiologically contaminated soils from a Department of Energy site using a biovented bioreactor system. Based on the success of their ex situ system, they conclude that both carbon dioxide and hydrocarbon levels should be monitored in situ as a means of assessing extent of petroleum removal.

As observed by Crapse et al., Ergas et al., who report findings of studies of natural attenuation of acid mine drainage, pH is a valuable parameter to gauge remediation activity in the field. Results of microcosm, field, and community analysis studies revealed the primary attenuating processes at this site to be dilution, silicate mineral dissolution, and Fe(III) and sulfate reduction. This combination of field- and laboratory-scale study has resulted in a better understanding of microbial diversity in acid mine drainage environments.

Use of molecular methods for assessing microbial identity, diversity, and activity in the environment is perhaps the most rapidly growing area of environmental study. Andria Costello reviews the state of these advances in molecular biological techniques with a focus on their broad application and utility in assessing the effectiveness of bioremediation throughout a field remediation process.

Modeling Approaches

As physicochemical and biological techniques have become more sophisticated and thus better able to overcome previously experienced shortcomings in linking field and flask scales, the authors in this section report advances in modeling techniques that are allowing better prediction and control of subsurface remediation operations. Sun and Glascoe provide a thorough description of the differences in the utility and output of analytical and numerical models for biodegradation and reactive transport and provide a broad listing of currently available codes of each type. With a focus on the tradeoffs between computational cost and resolution of simulated systems, these authors state the case of the value of each type of model in biodegradation and provide the outlook for future modeling approaches.

Gas transport mechanisms in the porous subsurface media, including viscous flow, Knudsen flow, diffusion, and thermal flow, and three models, Fick's first law, Stefan-Maxwell equations, and the dusty gas model, relevant to this application are the topic of Wa'il Abu-El-Sha'r's chapter. Here, he reviews each mechanism of flux and suitability of each model under various environmental conditions and emphasizes the importance of linking these gas transport models to both in situ and laboratory testing to provide the necessary gas transport parameters.

Brooks and Wood address the use of mass flux measurements of dense nonaqueous phase liquids (DNAPLs) as a remedial performance objective to overcome the weaknesses of current remediation methods in removing these persistent contaminants from source zone areas in the subsurface. These authors provide a review of four methods used to measure mass flux and complement this discussion with modeling approaches that enable prediction of mass flux changes during remediation. They report that three parameters–heterogeneity of the flow field, heterogeneity of the DNAPL distribution, and the correlation between these two heterogeneities–are key elements in assessing the benefits of any remediation technology in terms of mass flux.

References

1. U.S. Environmental Protection Agency. *Cleaning Up the Nation's Waste Sites: Markets and Technology Trends, 2004 Edition;* EPA 542-R-04-015; Office of Solid Waste and Emergency Response, U.S. EPA: Washington, D.C., September 2004.

2. U.S. Environmental Protection Agency. *Guidance for Evaluating the Technical Impracticability of Ground-Water Restoration;* EPA Directive 9234.2-25; Office of Solid Waste and Emergency Response, U.S. EPA: Washington, D.C., 1993.

3. U.S. Environmental Protection Agency. *Presumptive Response Strategy and Ex-Situ Treatment Technologies for Contaminated Ground Water at CERCLA Sites. Final Guidance;* EPA 542-R-96-023; Office of Solid Waste and Emergency Response, U.S. EPA: Washington, D.C., 1996.

4. Johnson, B.L.; DeRosa, C.T. The toxicologic hazard of Superfund hazardous waste sites. *Reviews on Environmental Health;* Freund Publishing House Ltd.: Atlanta, GA, 1997; Vol. 12, No. 4, pp 235-251.

5. Eller, G.; Kruger, M.; Frenzel, P. *FEMS Microbiology Ecology.* **2005**, *51*, 279-291.

Physicochemical Methods of Subsurface Remediation

Chapter 2

From Sequential Extraction to Transport Modeling: Monitored Natural Attenuation as a Remediation Approach for Inorganic Contaminants

Kimberly P. Crapse[1], Steven M. Serkiz[1], Adrian L. Pishko[1], Daniel I. Kaplan[1], Cindy M. Lee[2], and Anja Schank[2]

[1]Savannah River National Laboratory, Westinghouse Savannah River Company, Aiken, SC 29808
[2]Department of Environmental Engineering and Science, Clemson University, Clemson, SC 29634

To quantify metal natural attenuation processes in terms of environmental availability, sequential extraction experiments were carried out on subsurface soil samples impacted by a low pH, high sulfate, metals (Be, Ni, U, As) plume associated with the long-term operation of a coal plant at the Savannah River Site in South Carolina. Specific focus here is given to beryllium. Although nickel, uranium and arsenic were also analyzed, these results are not presented herein. Despite significant heterogeneity resulting from both natural and anthropogenic factors, sequential extraction results demonstrate that pH is a controlling factor in the prediction of the distribution of metal contaminants within the solid phases in soils at the site as well as the contaminant partitioning between the soil and the soil solution. Results for beryllium, in soils at the site as well as the contaminant partitioning between the soil and the soil solution. Results for beryllium, the most mobile metal evaluated, exhibit increasing attenuation along the plume flow path which corresponds to an increasing plume pH. These laboratory- and field-scale studies provide mechanistic information regarding partitioning of metals to soils at the site (one of the major attenuation mechanisms for the metals at the field site). Subsequently, these data have been used in the definition of the contaminant source terms and contaminant transport factors in risk modeling for the site.

Introduction

A large groundwater contaminant plume at the D-Area Expanded Operable Unit (DEXOU) at the Savannah River Site in South Carolina is characterized by high levels of acidity (low pH), metals, and sulfate. This plume is the result of the weathering of coal, coal ash, and coal spoils (materials that did not meet the specifications for combustion). The large plume and relatively low contaminant concentrations, therefore, make monitored natural attenuation (MNA) an attractive remediation strategy for inorganic contaminants at this site.

MNA (as defined by the U.S. EPA-OSWER Directive 9200.4-17P, April 1999 (*1*)) is the "reliance on natural attenuation processes (within the context of a carefully controlled and monitored site cleanup approach) to achieve site-specific remediation objectives within a time frame that is reasonable compared to that offered by other more active methods." Many natural processes in the soil act to mitigate transport and availability of contaminants and include a range of physical, chemical, and biological processes that reduce the mass, toxicity, mobility, volume, or concentration of contaminants in soil or ground water without the aid of human intervention. Implementation of MNA as a remediation strategy requires a mechanistic understanding of the natural attenuation processes occurring at the subject waste site. For inorganic contaminants, natural attenuation typically involves a decrease in metal concentration, toxicity and/or mobility. These natural processes that lead to natural attenuation of inorganics include: dilution, dispersion, sorption (including adsorption, absorption, and precipitation), and redox transformations.

To implement monitored natural attenuation for remediation of inorganic contaminants in soil, the site must first be evaluated and designated as an appropriate MNA site according to U.S. EPA protocols and guidelines using site-specific data. To lay the groundwork for establishing and validating MNA at the DEXOU, matched porewater/soil samples were collected along the plume flow path at the site. Soils were subjected to sequential extractions to provide site-specific metals availability data designed to elucidate the sorption processes contributing to natural attenuation of inorganic contaminants. Sequential extraction (*2,3*) is a desorption technique that has been useful for characterizing metal association with soils, as well as characteristics of reactive soil sorption surfaces. When considering the metal availability and transport of the contaminant, the porewater chemistry and the solid-phase metal speciation are important in defining the environmental availability of the contaminant (where environmental availability is defined by Amonette et al. (*4*) as "the ability of a soil to maintain an aqueous concentration of [contaminants] in the soil solution"). Therefore, the sequential extraction methodology needs to be matched to site conditions for maximum utility.

Sequential extraction and major ion chemistry data will ultimately be used as input for risk-based models used in the remedial decision-making process. The degree to which the risk-based models embody and accurately describe the major environmental processes that influence contaminant attenuation, therefore, has a direct effect on the appropriateness, cost effectiveness, and overall success of the site remediation.

Conceptual model

Contaminated runoff from the D-Area Coal Pile to the D-Area Coal Pile Runoff Basin has resulted in a low pH/metals/sulfate groundwater contaminant plume that emanates from the D-Area Coal Pile Runoff Basin. (See Figure 1 for site features and groundwater flow direction.) It was anticipated that environmental conditions at D Area would be favorable to the attenuation of metal contaminants. The groundwater plume becomes more anaerobic as it moves towards the Savannah River (i.e., from pyrite oxidation conditions near the Coal Pile Runoff Basin to sulfate-reducing conditions in the wetland), and pH rises due to acid-buffering from the site soils. Both of these conditions, more reducing redox potential (E_h) and increased pH values, are expected to attenuate metal contaminants by sorption and precipitation processes as the plume moves towards the Savannah River.

In addition to the primary source emanating from the vicinity of the D-Area Coal Pile Runoff Basin, a number of additional sources of metals contamination exist in D Area. A significant amount of sluiced ash overflowed from the D-Area Ash Basin in the 1970s into the wetland area west of the Ash Basin,

14

leaving metals-contaminated ash in the wetland. Additionally, from the west end of the Ash Basin as well as the D-Area Rubble Pit (DRP), there appear to be sources of low pH/sulfate/metals. The vertical stratification of the plume emanating from the vicinity of the Coal Pile Runoff Basin as well as multiple sources areas serve to confound interpretation of metal attenuation data strictly based on distance along the groundwater flow path from the primary source near the Coal Pile Runoff Basin. Nevertheless, the readily measurable parameter pH is thought to be a good predictor of metal attenuation at the site.

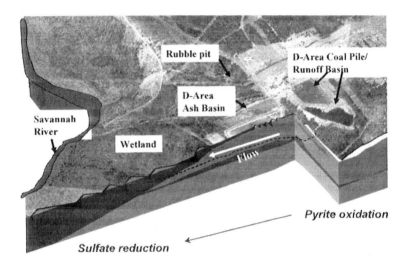

Figure 1. D-Area features.

Materials and Methods

Sampling

Soil cores were collected at multiple locations along the approximate flow path of the plume and also in the wetland area (Figure 2). Soil cores were collected at multiple subsurface depths (typically at two to three discrete intervals between 4 feet below ground surface to as deep as 53 feet below ground surface corresponding to elevations from 112 to 57 ft above sea level) for a given location, with the exception of the wetland locations (D2, D4, G10, H5, J6, K4), which were collected from the top 1 foot of ash deposition in the wetland area, and DAB86, which was collected from the bottom of the ash basin in perched water just above the clay layer below the ash basin (12 to 16 feet below ground surface). Soils were stored in sealed containers at 4°C.

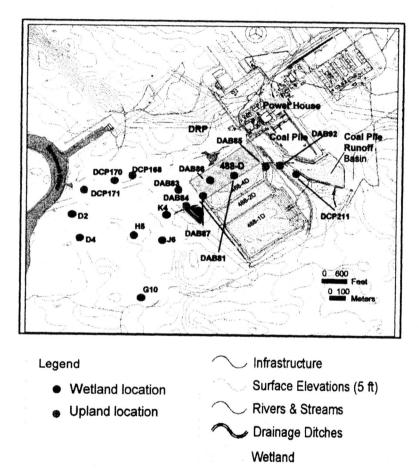

Legend

● Wetland location

⊕ Upland location

Infrastructure

Surface Elevations (5 ft)

Rivers & Streams

Drainage Ditches

Wetland

Figure 2. D-Area map showing sampling locations with respect to the Coal Pile, Coal Pile Runoff Basin, Ash Basin (488-D), and D-Area Rubble Pit (DRP).

Pore Water Analyses

Pore water was separated from a portion of the soil core within 12 hours of collection. The separation of soil and pore water was carried out using 50 mL centrifuge filter tubes, each fitted with a 20 mL capacity filter insert with either a 0.45 µm polypropylene membrane or 10 µm polypropylene mesh (Whatman®, Clifton, New Jersey). Typically, six tubes were filled with soil to the insert capacity and centrifuged at 7000 rpm for 10 minutes (0.45 µm filter) or at 1000 rpm for 10 minutes (10 µm filter). The insert was removed and the soil reserved for further analysis. E_h and pH for each sample were measured immediately following separation from the soil. Flow through pH and redox (E_h) electrodes with a Ag/AgCl flow through reference electrode (Microelectrodes, Inc., Bedford, New Hampshire) were used for these measurements. Pore water samples were analyzed for major ions by Inductively Coupled Plasma-Atomic Emission Spectroscopy (ICP-AES) (Jarrell-Ash 965, Franklin, Massachusetts, using a suite of 30 elements), trace metals by Inductively Coupled Plasma-Atomic Mass Spectroscopy (ICP-MS) (VG Plasma Quad III, Fisons Instruments, Danvers, Massachusetts), and for sulfate (SO_4^{2-}) using a Bran+Luebbe Auto Analyzer II Continuous Flow System (Norderstedt, Germany).

Soil Extraction/Digestion Methods

An eight-step sequential extraction procedure (Table I) was carried out on the subsurface soils collected, along with two other partial digestion methods: EPA method 3050b (hot nitric acid and hydrochloric acid) and a single-step extraction (corresponding to the amorphous iron oxide step (6th step) of the sequential extraction procedure (Table I)). Leachates were analyzed by ICP-AES and ICP-MS.

Results and Discussion

Methods Comparison

Only a fraction of the total metal concentration in a given soil is expected to be available for partitioning to the aqueous phase. Findley (5) has demonstrated that the bulk of naturally occurring trace metals in soil are only accessible under harsher extraction conditions (sequential extraction steps 7 and 8, Table I) that are not likely to represent conditions present in the environment. The available metals concentration in soil is important when considering both the source term as well as the partition coefficient derived from contaminated soil samples. By considering only the available metal concentration rather than the total metal concentration in soil, a large fraction of the naturally occurring trace metals pool is eliminated from consideration in fate and transport modeling of the site.

In this study, the sum of the metal mass in the first six sequential extraction steps ($\sum_{i=1}^{6} C_i$, where C_i is the metal concentration in soil for the ith extraction and i=1 through 6, Table I) has been operationally defined as the available metal concentration in soil. As a second method of defining the available metal concentration in soil, a single step extraction equivalent to step 6 of the sequential extraction procedure was found to yield equivalent concentrations compared to summing the first six sequential extraction steps in a series of split samples. This single step extraction and $\sum_{i=1}^{6} C_i$ are used interchangeably herein as the available metal concentration. The available metals concentration in soil can also be defined by the EPA 3050B extraction method (hot nitric acid). Data for all digestions are not presented in entirety herein but have been reported previously (6). Figure 3 presents a selection of representative data for the study site. Sequential extraction data for beryllium concentration in soil [stacked data: steps 7-8 (top); steps 1-6 (bottom)] is presented to the left of the EPA 3050B data for each location included. Upland locations, both near the source (low pH) and further downgradient (higher PH), as well as wetland data are presented. A background (unimpacted) wetland sample and a sample of ash from the ash basin were also included for comparison with the wetland sample. For the majority of the samples analyzed in this study, the EPA 3050B method overpredicts the available concentration of contaminant in soil relative to the sequential extraction method ($\sum_{i=1}^{6} C_i$) and the single step method (step 6)(data not shown).

For upland soils, in general, metal concentrations in soil (both available and total) are found to be higher at locations downgradient of the source area (impacted by the plume) corresponding to higher porewater pH [e.g., lower concentrations in soils with lower pH (nearer the source area)] (Figure 3). This observation is consistent with transport of beryllium from low-pH soils to higher-pH soils where beryllium is attenuated by increased sorption. The degree to which the downgradient soil has been impacted by the plume also affects the concentration of beryllium in the soil. The wetland is characterized by high total beryllium concentration in soil although a larger fraction of this total concentration in soil is defined as unavailable. Because the wetland was

Table I. Summary of Sequential Extraction Steps[1]

Availability	Fraction	Reagent	Description	Extraction Conditions	Targeted Phase
↑	1	Distilled deionized water		Tumble for 16 hours at rt[2]	Easily soluble salts and ions
	2	0.5 M calcium nitrate or magnesium chloride	neutral salt	Tumble for 16 hours at rt	Easily exchangeable ions on soil surfaces
	3	0.44 M acetic acid, 0.1 M calcium nitrate	weak acid with neutral salt	Tumble for 8 hours at rt	Carbonate minerals, acid exchangeable metals on the soil surfaces
	4	0.01 M hydroxylamine-hydrochloride, 0.1 M nitric acid	weak reducing agent	Tumble for 0.5 hours at rt	Manganese oxides
	5	0.1M sodium pyrophosphate (SP) or hydrogen peroxide (HP)	oxidizing agent	Tumble 24 hours at rt/SP or 85°C for 5 hours /HP	Organic matter
	6	0.175 M ammonium oxalate, 0.1 M oxalic acid	buffered mild reducing agent	Tumble 4 hours in darkness at rt	Amorphous iron oxides
	7	0.15 M sodium citrate, 0.05 M citric acid, 25 g/L sodium dithionite	buffered strong reducing agent	Shake for 0.5 hours in water bath at 50°C	Crystalline iron oxides
	8	48% hydrofluoric acid, aqua regia	Strong corrosive	Microwave digestion	all remaining solids

[1]based on Miller (2) and Tessier (3)
[2]rt = room temperature

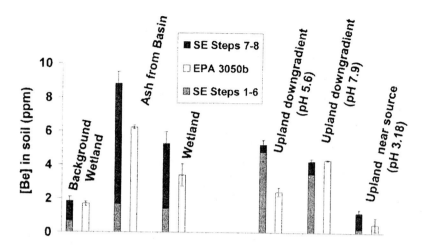

Figure 3. Beryllium concentration in soil for selected locations as determined by sequential extraction and EPA 3050B methods.

contaminated with ash material, it is difficult to determine whether the higher concentrations are due to a natural shift in attenuation mechanism from upland to wetland soils or from physical transport of beryllium-containing ash into the wetlands. More likely, the observed results are due to the physical transport of ash material from the basin, which is also characterized by high concentrations of beryllium in the unavailable fractions.

Beryllium Availability in Soil

The percentage of the total beryllium concentration in soil potentially available to the mobile aqueous phase was estimated using sequential extraction data and eq 1:

$$\% \ Available = \left(\frac{\sum_{i=1}^{6} C_i}{\sum_{i=1}^{8} C_i} \right) \times 100 \qquad (1)$$

where C_i is the metal concentration in soil determined after the ith extraction step and i = 1 through 6 (available beryllium), or I = 1 through 8 (total beryllium) respectively (Table I) as previously described.

Sequential extraction availability profiles are useful not only in considering availability in terms of the source term definition (i.e., what percentage of the metal pool is available) but also as indicative of distinctive attenuation mechanisms. For beryllium, typically, lower total metal concentrations in soil were found in areas most impacted (lowest pH). These soils also exhibited lower percentages of the total beryllium in the available fraction suggesting that the more available beryllium had been preferentially leached by the low-pH plume. Higher soil concentrations with a higher percentage of beryllium in the available fraction were found to correlate with attenuation of acidity (higher pH). The percentage of available metal in soil was highly correlated to pH (the regression is significant at 1% level of probability $p < 0.01$ with 18 degrees of freedom) (Figure 4). At higher pH the variable surface charge becomes increasingly negative resulting in a corresponding increase in sorption of cationic beryllium. Alternatively or in addition to the increasingly negative charge at higher pH values, formation of Fe-(oxy)hydroxides at higher pH values are also expected to increase the sorption of cations in this system.

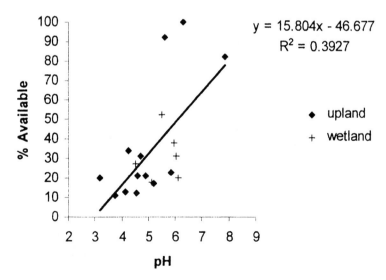

Figure 4. Percentage of the available beryllium concentration in soil as a function of pH.

Distribution Coefficient

Typically, a linear partitioning coefficient (K_d) used in groundwater models is defined as shown by eq 2:

$$K_d \; (mL/g) = \frac{\text{Contaminant concentration in soil (mg/kg)}}{\text{Contaminant concentration in the solution contacting the soil (mg/L)}} \qquad (2)$$

K_d values are most often obtained from published literature, generated in lab-scale batch sorption experiments or generated from contaminated soils in desorption experiments. Variations of three to four orders of magnitude are not uncommon for K_d values from the literature or even from the same waste site where there are large geochemical gradients.

Because only the available metal concentration would be expected to partition to the aqueous phase, K_d values were calculated based on the available concentration of beryllium in soil. The log of K_d is significantly correlated to pH (Figure 5). K_d values are quite large indicating the high attenuation capacity

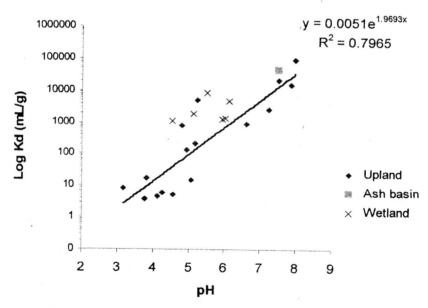

Figure 5. Log K_d as a function of pH (upland soils only included in regression shown).

of the D-Area soil for beryllium. At the D-Area site, K_d values vary over four orders of magnitude, clearly a single linear K_d approach would not be appropriate for modeling beryllium transport. Alternatively, a model based on K_d as a function of pH is being evaluated for this site (7).

Conclusions

Beryllium attenuation in soil was characterized at the field site impacted by coal plant operations. Despite natural and anthropogenic heterogeneities at the site and the inherent difficulties associated with the interpretation of field data, the attenuation of beryllium was highly correlated with the readily measurable parameter pH. Higher soil concentrations of beryllium with a higher percentage of the metal in the available fraction corresponded to soils with higher pH, indicating beryllium attenuation. Sequential extraction data were used to define the percentage of the beryllium concentration in soil that was available as well as the partitioning coefficients associated with this available fraction. Both of these site-specific parameters correlated with pH and will provide input for groundwater modeling efforts at the site.

Acknowledgements

This work was funded by the U. S. Department of Energy under contract DE-AC09-96SR18500.

References

1. U.S. Environmental Protection Agency, Office of Solid Waste and Emergency Response, April 1999, Use of Monitored Natural Attenuation at Superfund, RCRA Corrective Action, and Underground Storage Tank Sites. OSWER Directive 9200.4-17P, 1999.
2. Miller, W.; Martens, D.; Zelazny, L. *J. Soil Sci. Soc. of Am.* **1986**, *50*, 598-601.
3. Tessier, A.; Campbell, P.G.C.; and Bisson, M. *Anal. Chem.,* **1979**, *51*, 844-851.
4. Amonette, J. E.; Holdren, G. R.; Krupa, K. M.; Lindenmeier, C. W. Assessing the Environmental Availability of Uranium in Soils and Sediments. U.S. Nuclear Regulatory Commission report NUREG/CR-6232, 1994.

5. Findley, M. M.S. Thesis, Clemson University, Clemson, SC, 1998.
6. Crapse, K. P.; Serkiz, S. M.; Pishko, A. L.; McKinsey, P. C.; Brigmon, R. L.; Shine, E. P.; Fliermans, C.; Knox, A. S. Monitored Natural Attenuation of Inorganic Contaminants Treatability Study Final Report. WSRC-TR-2004-00124, 2004.
7. Brewer, K. E.; Sochor, C. S. Flow and Transport Modeling for D Area Groundwater (U) WSRC-RP-2002-4166, 2002.

Chapter 3

Effects of Methanol on Remediation of Soil Contaminated with Toxaphene

Xiaosong Chen and Clayton J. Clark II

Department of Civil and Coastal Engineering, University of Florida, Gainesville, FL 32611

Due to its persistence in nature and harmful effects on biota, toxaphene, a once prevalently used pesticide, has been a major concern for health and environmental officials. The study of toxaphene and methods to remediate it effectively and efficiently continues to be a major issue for environmental engineers and scientists alike. This research explored the potential for cosolvent flushing to remove toxaphene contaminants from soil using methanol-water solutions. The research revealed that the addition of methanol to water increased the solubility of toxaphene as it decreased its sorption to the soil. The first-order bicontinuum model was found to effectively simulate the non-equilibrium sorption process in the miscible displacement of toxaphene from a soil column. The application of cosolvent flushing reduced the time and solution volume required to elute the hydrophobic organic toxaphene from the soil. The experimental data agreed with the theoretical log-linear relationships between the sorption constant K_p and the cosolvent fraction f_c, and the reverse first-order rate k_2 and the cosolvent fraction f_c. A log-log linear relationship was found to exist between the k_2 value and the K_p value. The linear correlation between k_2 and the cosolvent fraction f_c was estimated successfully based on the literature data and equations.

Introduction

Toxaphene, a polychloroterpene material, was developed in 1947 (*1*) and was used mainly as a pesticide for many southern crops, especially cotton. Generally, toxaphene is an amber, waxy organic solid with odor similar to turpentine (*2*) and is comprised of at least 3200 closely related polychlorinated compounds, the majority of which are chlorobornanes ($C_{10}H_{16-n}Cl_n$) and chlorocamphenes ($C_{10}H_{18-n}Cl_n$), where n = 5–12. Toxaphene is harmful if contacted with skin, toxic if swallowed, and an irritant to the respiratory system (*2*). It has been reported to cause long-term adverse effects in the aquatic environment (*2*), and is easily metabolized once ingested by birds and mammals.

A dense non-aqueous phase liquid (DNAPL), toxaphene is not leached appreciably into groundwater and cannot be removed significantly by runoff unless adsorbed by sediments due to its low solubility in water (3 ppm). In accordance with this, toxaphene has been found to be strongly adsorbed to sediments (Log K_{ow}= 3.3) and has been proven to be difficult to break down in aerobic environments (*3*). When released to the soil, toxaphene has been noted to persist for long periods of time, up to 14 years (*1*), and thus, it is more likely to be found in soil or sediment at the bottom of lakes and streams. By-products from toxaphene have also been shown to be potentially detrimental as well (*4*). Due to its harmful nature and ability to persist in the environment, it has become necessary to study the degradation of toxaphene to decrease its presence in the environment or remove it altogether.

The process of cosolvent flushing has been found to provide rapid mass removal of NAPL at sites with moderate to good permeability (*5*). It involves pumping a mixture of water and one or more solvents through a contaminated zone to remove NAPL by dissolution and/or mobilization. Alcohols such as methanol have been increasingly used in the remediation of subsurface contamination involving chlorinated organics, such as PCE and TCE, which can form sparingly soluble chlorinated organic compounds (*5*). Alcohol cosolvents increase NAPL solubility in water by lowering the NAPL-water interfacial tension in the fluid throughout the subsurface in-situ flood. This research seeks to examine the potential effectiveness of cosolvent methanol in removing toxaphene from soil.

The objectives of this research included: 1) examination of toxaphene solubility in water for different fractions of methanol in solution; 2) derivation of parameters of a non-equilibrium bi-continuum model to describe increased solubility of toxaphene in solution; 3) derivation of the relationship between various model parameters describing the methanol flushing of toxaphene contaminated soil.

Background

Solubility of Organic Compounds

Enhancement of solubilization of compounds with low aqueous solubility is described mathematically using a cosolvency power (6). In calculation of the cosolvency power value, a log-linear equation was found to describe the phenomenon of the exponential increase of the solubility for nonpolar organic solute as the cosolvent fraction increased (6): $\log(S_{s,m}) = f_c \log(S_{s,c}) + (1-f_c)\log(S_{s,w})$, where $S_{s,m}$ is the log solubility (moles/L) of a solute in a mixed solvent system; $S_{s,w}$ is the weighted sum of the solute log solubility in pure water; and $S_{s,c}$, is the solubility of the solute in pure cosolvent. For a given cosolvent-water system, the difference between log solubilities of the solute in pure cosolvent and pure water was defined as the cosolvency power, σ, of this cosolvent (6) as $\sigma = \log(S_c/S_w)$, where S_c and S_w are molar solubilities of solute in pure organic cosolvent and in pure water, respectively (6, 7). The values of σ can be positive, near-zero, or negative, depending on the relative polarity of water, solute, and cosolvent (7). Solute solubility in a mixed solvent system can also be expressed in a theoretical log-linear form (7) as $\log(S_{s,m}) = \log(S_{s,w}) + \Sigma \sigma_i f_{ci}$, where f_{ci} is the volume fraction of ith cosolvent; σ_i is the cosolvency power for solute in pure i^{th} cosolvent. Similar to Raoult's law, in which the partial pressure of a component in a mixed liquid is the product of its mole fraction and its vapor pressure (8), the log-linear relationship of solubility is based on an assumption of the ideality of the solutions in which all components of the mixture behave identically.

However, mixed solutions are generally not ideal, and the log-linear model shows some deviations compared to the experimental data (9, 10). To quantify the deviation caused by the nonideality of the solvent mixture, the above equation for σ was modified as follows: $\log(S_{s,m}) = \log(S_{s,w}) + \Sigma \beta_i \sigma_i f_{ci}$, where β_i is the empirical coefficient used to account for water-cosolvent interactions. For methanol, the interaction of water-cosolvent interaction is negligible, and β is always to be assumed as 1 (16). However, because β is not independent of the change of cosolvent fraction, the application of the equation for the solubility prediction should be investigated carefully over a specific range of cosolvent fractions.

To compensate for nonideality, a computational modification, known as UNIFAC (UNIQUAC functional-group activity coefficients), has been employed (11-15). The basic assumption of UNIFAC is that a physical property of a fluid is due to the sum of contributions made by the molecule's functional groups (13-15). The UNIFAC model allows the necessary parameters to be estimated from the number and type of functional groups that comprise the chemical species. Li (13) further provided an approach to accurately predict cosolvency power, by extending the log-linear model with the activity coefficients of the system components. Pinal et al. (15) proposed that a term 2.303 $\Sigma(f_l \log \gamma_i)$, which is the analogue to $\Sigma(X_i \ln \gamma_i)$, be added to the simple log-linear model to account for the effect of the solvent nonideality. Li (13) extended this equation to four equations (eqns 1-4) with volume fraction f_c or mole fraction x of the cosolvent in the mixture, and logarithms of the activity coefficient terms that are 10-based or e-based:

$$\log S_{s,m} = \log S_{s,w} + \sigma f_c + \log \gamma_w + f_c \log\left(\frac{\gamma_c}{\gamma_w}\right) \tag{1}$$

$$\log S_{s,m} = \log S_{s,w} + \sigma f + \log \gamma_w + x \log\left(\frac{\gamma_c}{\gamma_w}\right) \tag{2}$$

$$\log S_{s,m} = \log S_{s,w} + \sigma f_c + \ln \gamma_w + f_c \ln\left(\frac{\gamma_c}{\gamma_w}\right) \tag{3}$$

$$\log S_{s,m} = \log S_{s,w} + \sigma f_c + \ln \gamma_w + x \ln\left(\frac{\gamma_c}{\gamma_w}\right) \tag{4}$$

Cosolvent Flushing Process

For equilibrium processes, there is an assumption that the sorption-desorption process is so rapid that it is instantaneous; that is, the point-wise equilibrium exists between the solution and sorbent phase. However, in all practicality, the sorption-desorption processes are not instantaneous. To study the sorption and transport of toxaphene in a miscible displacement column experiment, a non-equilibrium bi-continuum sorption model used successfully by previous researchers (17,18) was employed. In this model, the sorption process is divided among two types of sites, S_1, the instantaneous domain (g/g) and S_2, the first order, rate-limited domain (g/g), as shown in eqs 5 and 6.

$$S_1 = FK_p C \tag{5}$$

$$\frac{dS_2}{dt} = k_1 S_1 - k_2 S_2 \tag{6}$$

where F is the ratio of the instantaneous sorption to the total sorption domain; K_p is the equilibrium sorption constant (cm^3/g); k_1 is the forward first-order rate constant (hr^{-1}); and k_2 is the reverse first-order rate constant (hr^{-1}). The governing equation is:

$$\frac{\partial C'}{\partial p} + (\beta R - 1)\frac{\partial C'}{\partial p} + (1-\beta)R\frac{\partial S'}{\partial p} = \left(\frac{1}{P}\right)\frac{\partial^2 C'}{\partial X^2} - \frac{\partial C'}{\partial X} \tag{7}$$

$$(1-\beta)R\frac{\partial S'}{\partial p} = \omega(C' - S') \tag{8}$$

where C' $=C/C_0$, distance X = x/l, and time p = vt/l are dimensionless; and P is Peclet number = vl/D. Equations for calculation of other parameters include the following:

$$S' = \frac{S_2}{(1-F)K_p C_0} \tag{9}$$

$$R = 1 + \left(\frac{\rho}{\theta}\right)K_p \tag{10}$$

$$\beta = \frac{\left[1 + F\left(\frac{\rho}{\theta}\right)K_p\right]}{R} \tag{11}$$

$$\omega = \frac{k_2(1-\beta)Rl}{v} \tag{12}$$

where D is the dispersion coefficient (cm^2/hr); v is the average pore-water velocity (cm/hr); l is the column length (cm); ρ is the bulk density (g/cm^3); and θ is the volumetric soil-water content. In this model, five parameters must be calculated or estimated, T_0, P, R, β, ω. In this study, the value of the input pulse T_0 (hr) was set and measured in experiments. R, the retardation factor, was calculated from moment analysis (18); P, Peclet number, was derived from a IPA tracer test; and the values of β and ω were estimated from the program

CXTFIT 2.1 (USDA, ARS, Riverside, California). β is the fraction of the instantaneous domain retardation in the total sorption retardation. ω is a measure of the mass transfer limitations which represents the ratio between hydrodynamic residence time and the characteristic time of sorption.

Materials and Methods

Materials

The selection of methanol as the cosolvent solution was due to its non-toxicity, biodegradability, economic efficiency, and availability. The column was packed with Eustis #3 fine sand from Gainesville, Florida with 0.39% organic carbon and 96% sand content. The experiments were conducted in 2.5 cm ID * 15 cm length glass column (Kontes Glass Co., Vineland, NJ) fitted with Teflon o-rings and Teflon end pieces. The length of the column was adjusted to 5 cm which was employed by other researchers (19, 20). All tubing, fitting, and valves in contact with the organic fluids were constructed with Teflon. The bulk density of this sandy column was 1.7 g cm^{-3}, and the porosity was 0.39. Isopropyl alcohol (IPA) was used as non-reactive tracer within the column.

Solubility

Initially, 5 mg of pure toxaphene was placed into the 5 mL empty vials with a Teflon-coated, septum-lined caps. Different methanol fractions of 0%, 10%, 20%, 30%, 40%, 50%, 60%, and 75% were used in methanol-water solutions that were added to the vials leaving no headspace. Volumes of methanol and water were measured separately and combined to avoid the volume change due to the volume shrinkage during mixture. The solutions were equilibrated on a rotator for 48 hrs at room temperature. Preliminary tests showed that this time was sufficient for equilibrium of the batch systems to be reached. If the micelle formed in the solution, the vials were centrifuged at 3000 rpm for 25 mins.

Gas chromatography (GC-17A, Shimadzu, Columbia, MD) was used for analysis of toxaphene in the hexane-extracted aqueous phase. The total area approach was used for measuring the toxaphene present in the hexane-extracted aqueous phase due to the chromatographic complexity observed with this compound (21-23). The operating conditions were as follows: column,

DB-5 with 30 m length, 0.32 mm inner diameter, and 0.25 µm film thickness; carrier gas, H_2; injector temperature, 220°C; detector temperature, 300°C; and the column temperature program: an initial temperature of 100°C to 150°C at 30°C/min was held for 2 min and then temperature of 150°C ramped to 300°C at 3°C/min held 5 min.

Column Test for Miscible Displacement

The miscible displacement experiment was similar to that used by Brusseau et al. (19). The column packed with sandy soil was initially saturated with water by pumping 0.01 N $CaCl_2$ DI water with an upward flow rate of 0.5 mL/min. All solutions were prepared with a 0.01 N $CaCl_2$ matrix.

IPA was employed as a non-reactive tracer with 2.3 pore volume pulse input. The tracer experiments were performed before and after the cosolvent flushing. The results were not significantly different, suggesting that cosolvent flushing had little effect on the retention capacity of the soil column during the experiment periods.

The cosolvent system consisted of methanol solutions with fractions of 30%, 40%, 50%, 60%, and 75% methanol. The initial concentrations, C_0, for every fraction of cosolvent solution were within 50% to 80% of the solubility limit of the cosolvent solutions to satisfy the linear sorption requirement. Results indicated that breakthrough curves using C/C_0 were almost the same, in the range of C_0 from 40% to 80% of solubility. Approximately 2.3 pore-volumes of solution containing toxaphene were injected into the column with an upward pulse input flow rate of 1 mL/min (9.57 cm/hr), followed by the flushing of the cosolvent solution with upward flow rate of lmL/min. A switching valve was used to facilitate switching between the solution containing toxaphene and the cosolvent solution without toxaphene. The column effluent was collected every 1 mi with an automatic collector. The collection time for the 30% methanol solution was 240 min (25 pore volume), for the 40% methanol solution, 180 min (18.8 pore volume), and, for the other solutions, 120 min (12.5 pore volume). Mass recoveries were >95% for all experiments.

Results and Discussion

Solubility of Toxaphene

Research was conducted to analyze how the addition of methanol as a cosolvent affected the aqueous solubility of the pesticide toxaphene. A

32

statistical analysis comparing the log-linear and the extended log-linear models including the UNIFAC method (13) is shown in Figure 1. In this analysis, the percentage difference between the estimated log of toxaphene solubility and the experimental value was calculated and compared using the expression %difference $[(\log S_{s,expt} - \log S_{s,est})/\log S_{s,expt}]*100$, and plotted as a function of the fraction (f_c) of methanol present. As seen in Figure 1, the extended models produced higher average percentage errors than the simple log-linear model. This phenomenon may be attributed to the near ideality of the mixture of water and the highly polar methanol. Previous research has shown that log-linear solubilization has been shown to prevail in mixtures such as these (13).

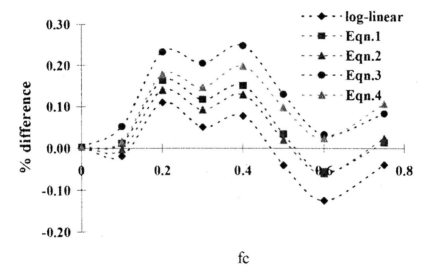

Figure 1. Comparison of the experimental data of toxaphene solubility in methanol to estimated values using equations and data provided in (9) and (13).

Experimental results also produced a log-linear plot for the methanol-water system as shown in Figure 2. This plot indicated that the cosolvency power for toxaphene solubilized by methanol was 3.43, which mirrored the data estimated from theoretical log-linear equation and the extended log-linear eqns 1-4.

The log-linear model can be modified with the solvent-water interaction constant β(*16*). Using the estimated cosolvency power of methanol, the values

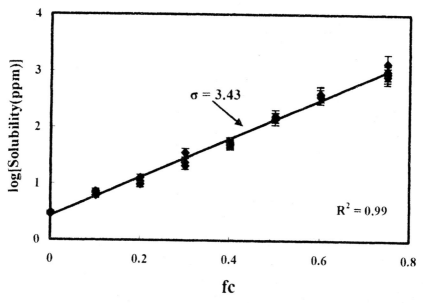

Figure 2. Cosolvency power from the log-linear regression of the log of toxaphene solubility in solution as a function of methanol fraction.

of β were nearly 1 for the various methanol fractions used, which validated the assumption that the solvent-water interaction is negligible for the methanol-water system.

Results for Column Test

Breakthrough curves (BTCs) from the methanol flushing of soil in a column contaminated with toxaphene are shown in Figure 3. The BTCs for the toxaphene flushed from the soil by solutions that contained different fractions of methanol exhibited some asymmetrical properties and had some degree of "tailing" which is characteristic of nonequilibrium caused by sorption of toxaphene to the soil. The BTC that used the IPA tracer was symmetrical and

34

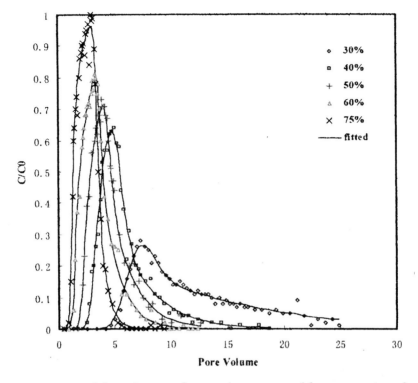

Figure 3. Breakthrough curves for toxaphene extracted from contaminated soil column by methanol/water cosolvent solution.

can be described by the advective-dispersive local equilibrium solute transport model:

$$\frac{dC'}{dp} = \left(\frac{1}{P}\right)\frac{d^2C'}{dX} - \frac{dC'}{dX} \tag{13}$$

In the present research, the column flow characteristics included a fluid velocity of 31.33 cm/hr, a diffusion coefficient of 2.53 cm^2/hr, and a dimensionless Peclet number, acquired from fitting the model to the data, of 61.9. The Peclet number measured in these tracer experiments was comparable to other research with similar column systems (*20*) and was used for all subsequent cosolvent flushing experimentation.

A one-dimensional flow model based on the bicontinuum approach was employed for further analysis of the methanol column flushing and required five parameters (T_0, P, R, β, and ω). T_0 is known as the pore volume and was set at 2.3 pore volumes, and the values for R and thus K_p were obtained by moment analysis (18):

$$R = M_1 - 0.5T_0 \qquad (14)$$

where

$$M_1 = \frac{\int_0^\infty C'p\,dp}{\int_0^\infty C'\,dp} \qquad (15)$$

where M_1 is the first moment and retardation value was in the situation of a non-Dirac input pulse (with finite T_0 of 2.30). The value of K_p was calculated from the R value with eqn 10.

The values of β and ω were gathered from fitting the data using the CXTFIT 2.0 computer modeling software. β is the fraction of the instantaneous domain retardation in the total sorption retardation, and ω is the Damkholer number which is a measure of the mass transfer limitations and represents the ratio between hydrodynamic residence time and the characteristic time of sorption. Then from eqn 12, the value of k_2, the reverse first-order rate of concentration, was calculated. All modeling results are shown in Table I.

Table I. Parameter Values for Binary Cosolvent System

f_c	30%	40%	50%	60%	75%
R	12.7	4.90	3.55	2.40	1.49
β	0.502 ± 0.007	0.701 ± 0.013	0.684 ± 0.110	0.631 ± 0.170	0.786 ± 0.300
ω	1.360 ± 0.060	0.706 ± 0.087	0.810 ± 0.083	0.958 ± 0.113	0.627 ± 0.214
K_p (cm^3g^{-1})	2.684	0.895	0.585	0.321	0.112
K_2 (hr^{-1})	1.347 ± 0.093	3.019 ± 0.52	4.524 ± 0.594	6.778 ± 1.048	12.321 ± 6.433
F	0.459 ± 0.008	0.624 ± 0.016	0.560 ± 0.150	0.367 ± 0.270	0.349 ± 0.091

Relationship between Sorption and Cosolvent Fraction

In the flushing of the toxaphene-contaminated soil in the column, results showed that increased methanol content decreased the retardation of the toxaphene elution from the soil. This decrease can be attributed to the increase of the aqueous solubility and decrease of sorption to the soil of the toxaphene. The sorption constant, K_p, plotted as a function of methanol fraction f_c in a log linear manner is shown in Figure 4. The slope of the regression is the collective term - $\alpha \beta \sigma$, where α is a constant representing the solvent-sorbent interactions, β is a constant representing the water-cosolvent interactions, and σ is the cosolvency power. The value of $\beta \sigma$ for methanol was obtained from the solubility test was 3.50, and the slope for methanol, $\alpha \beta \sigma$, was 2.91. Thus, the value of a was 0.83, meaning the solvent-sorbent interactions in this situation cannot be ignored. This α value was also consistent with those reported for anthracene and naphthalene sorption by methanol (*19*).

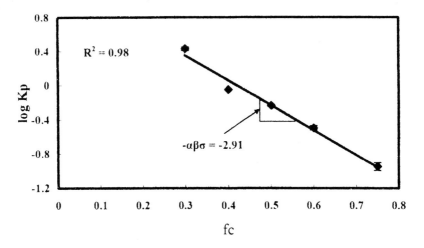

Figure 4. logKp-fc for methanol/water cosolvent system.

Reaction Rate as a Function of Sorption

The value of k_2 and K_p were seen to be inversely related based on experimental results (Table I). Other researchers have also explored the relationship between these two factors. Augustijn et al. (24) developed a log k-log K_{oc} relationship using data for various soil types, log k = 3.53-0.97logK$_{oc}$, where K_{oc} is the equilibrium sorption coefficient for organic carbon. Augustijn (25) adjusted this equation according to the "aging effects" on the mass transfer rates, log k = 1.74-0.69logK$_{oc}$. Brusseau et al. (19) studied the k_2 and K_p relationship using Eustis fine sand and methanol solution with naphthalene and anthracene as solute, and obtained the regression relationship, log k = 0.79-0.61logK$_{oc}$. Using the same column apparatus and various polynuclear aromatic hydrocarbons (PAHs) as solutes, Bouchard (26) developed the relationship between k_2 and K_p as log k = 0.47-0.91logK$_{oc}$.

Based on the results of these researchers, the value of τ (slope of log k_2 – log K_p) was in the range among 0.61 to 0.97. In the present research, a log-linear plot shown in Figure 5, yielded a value of 0.71 for τ corresponding to methanol cosolvent flushing of the toxaphene contaminated soil. Therefore, the results from the present research were consistent and found to be within the range of those reported in the literature. Based on the results of the present research, the following relationship was produced, log k_2 = 0.45-0.71logK$_{oc}$. The y-intercept is the value of logk$_{2,w}$, the reverse sorption rate constant in a pure water system. From extrapolation of this plot, the value of $k_{2,w}$ for toxaphene was found to be approximately 3 hr^{-1}.

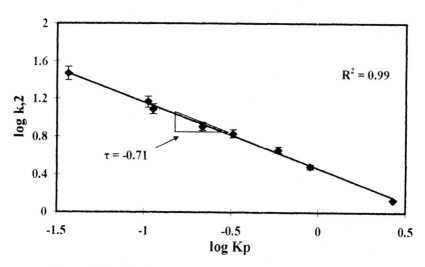

Figure 5. log k2– logKp for methanol/water cosolvent system.

Reaction Rate as a Function of Cosolvent Fraction

A log-linear relationship between the reaction rate and the methanol fraction in solution was determined from experimentally derived data. The plot of log k_2 as a function of methanol fraction (f_c) is shown in Figure 6 and can be expressed as log $k_2 = -0.47 + 2.051 \log K_{oc}$. The results indicated that the reverse rate constant, k_2, increases with an increase in methanol fraction in solution. This phenomenon may be attributed to the fact that the increase in cosolvent can lead to a larger diffusion coefficient and, therefore, increase the first-order rate constant. The value of the slope, Φ, comprises four terms: τ, the slope of the $\log k_2 - \log K_p$; α, the solvent-sorbent interaction constant; β, the solvent-solute interaction constant; and σ, the cosolvency power.

The relationship of log k_2 and f_c was also estimated based on theoretical information. The τ value of 0.61 was reported by Brusseau et al. (*19*) through studying several HOCs in aquaeous system. The cosolvency power σ for methanol was estimated as 3.50, as discussed in the solubility section. The value of β was assumed to be 1 and the value for α was estimated as 0.9 according to report of Brusseau et al. (*19*). Based on these assumptions, the value of the slope Φ could be predicted as 2.10 for methanol, which was similar to the value of 2.05 gathered from the results of the present research. Due to the complication of the toxaphene chemical structure, its referenced value of log K_{oc} is presented as a wide range from 2.47 to 5.00 (*27*). The value of $K_{p,w} = K_{oc}(OC)$, and the OC for Eustis find sand is 0.74%. Thus, the $K_{p,w}$ for toxaphene was estimated within the range from 1.42 to 470. According Brusseau's research (*19*), the value of $\log k_{2,w}$ for toxaphene should be between the values of -0.8 and 0.7. The plot of log k_2- f_c from the methanol cosolvent flushing column test is shown in Figure 6, and the calculated the value of log $k_{2,w}$ was -0.41, within the range predicted by the literature.

Sorptive Domain as a Function of Cosolvent Fraction

The parameter of F describes the distribution of sorption between instantaneous-to-total sorptive domains and may be calculated from the value of β according to eqn 11. The relationship of F and f_c for methanol indicates the F value would decrease with an increase of methanol fraction within the range of the 40% to 75% colsolvent fraction. Brusseau et al. (*19*) observed the same phenomenon for up to 20% cosolvent. Augustijn et al. (*25*) proposed the equation for this trend, $F^m = F^w - (F^w - 0.01) (f_c - 0.2)/0.8$ for $0.2 < fc < 1$, where the superscripts w and m indicate the water and cosolvent mixture, respectively.

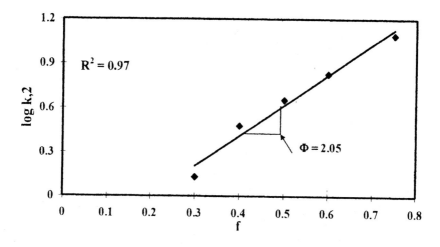

Figure 6. log $k_2 - fc$ for methanol/water cosolvent system.

Using this equation and the experimental data of methanol flushing, the F_w was calculated to be approximately 0.8.

This phenomenon may be explained by the swelling polymer property (*28*). At the first stage of swelling, the addition of cosolvent solution to the soil would enhance the thickness of the matrix more quickly than the increase of the surface, which results in the relative increase of thickness and reduction of surface area to volume ratio. Therefore, with the increasing of the cosolvent fraction within a certain range, the F value would be reduced with the decrease of instantaneous surface-to-internal matrix. However, the experiment results showed that under the fraction of 40% methanol, the F value was not decreased with the increase of fraction, which may be due to some possible critical or threshold level of cosolvent to show appreciable swelling effect (*19*).

Conclusions

Aqueous solubility of toxaphene in methanol-water solutions was expressed as the cosolvency power, measured by log-linear regression analysis. This cosolvency power for toxaphene in a methanol-water binary system was 3.43, which paralleled theoretical data estimated from the models in the literature.

The first-order bi-continuum model was found to effectively simulate the non-equilibrium sorption process in the miscible displacement of toxaphene from a soil column. The application of cosolvent flushing was shown to reduce the time required to remove toxaphene from soil by increasing its solubility in water used to flush the system. The higher the fraction of methanol in solution, the lower the retardation factor and sorption of toxaphene to the soil.

The experimental data mirrored values in the literature for the log-linear relationship of sorption constant K_p as a function of cosolvent fractions. The results also showed a log-linear relationship between the reverse first-order rate k_2 and the cosolvent fraction. A log-log linear relationship was also found between the k_2 value and the K_p value. Futhermore, experimental k_2 values compared favorably with values estimated from literature data and theoretical equations.

Overall, methanol in water was found to act as an effective agent for removal of toxaphene from soil. If this process is scaled up to potential field application, contaminant removal efficiency with cosolvents such as methanol may lead to more cost effective remediation of field sites contaminated with toxaphene and other complex pesticides.

References

1. U.S. EPA Environmental Protection Agency. Public Health Goal for Chemicals in Drinking Water, 2003. http://www.epa.gov/safewater/ mcl.html. [Date last accessed: 11/28/05.]
2. Agency for Toxic Substances and Disease Registry. Toxicological Profile for Toxaphene. Atlanta, GA: ATSDR, Public Health Service, U.S. Department of Health and Human Services. **1996**; pp. 96.
3. Nash, R.G.; Woolson, E.A. *Science*. **1967**, *157*, 924-927.
4. Parr, J.F.; Smith, S. 1976. *Sol. Science*. **1976**, *121, 51-57*.
5. Jawitz, J.W.; Annable, M.D.; Rao, P.S.C.; Rhue, R.D. *Environ. Sci. Technol.* **1998**, *32*, 523.
6. Yalkowsky, S.H.; Roseman, T. *J. Pharm. Sci.* **1980**, *69*, 912-922.
7. Li, A; Yalkowsky, S.H. *Ind. Eng. Chem. Res.* **1998**, *37*, 4470-4475.
8. Prausnitz, J.M.; Lichtenthaler, R.N.; deAzevedo, E.G. *Molecular Thermodynamics of Fluid-Phase Equilibra;* 2nd ed., Prentice-Hall Inc.: Englewood Cliffs, N.J., **1986**; pp. 168.
9. Morris, K.R.; Abramowitz, R.; Pinal, R.; Davis, P.; Yalkowsky, S.H. *Chemosphere*. **1988**, *17*, 285-298.
10. Li, A.; Andren, A.W. *Environmental Science and Technology*. **1994**, *28*, 47-52.
11. van Genuchten, M.Th. *J. Hydrol.* **1981**, *49*, 213-233.

12. Powers, S.E.; Hunt, C.S.; Heermann, S.E.; Corseuil, H.X.; Rice, D.; Alvarez, P.J.J. *Critical Review in Environmental Science and Technology.* **2001**, *31,* 79-123.

13. Li, A. *Ind. Eng. Chem. Res.* **2001**, *40*, 5029-5035.

14. Gupte, P.A.; Danner, R.P. *Ind. Eng. Chem. Res.* **1987**, *26,* 2036-2042.

15. Pinal, R.; Rao, P.S.C.; Lee, L.S.; Cline, P.V. *Environmental Science and Technology* **1990**, *24,* 639-647.

16. Rao, P.; Lee, L.; Pinal, R. *Environmental Science and Technology* **1990**, *24,* 647-654.

17. Nkedl-Kizza, P.; Rao, P.S.C.; Hornsb, A.G. *Environmental Science and Technology.* **1987**, *21,* 1107-1111

18. Brusseau, M.L.; Jessup, R.E.; Rao, P.S.C. *Environmental Science and Technology.* **1990**, *24,* 727-735.

19. Brusseau, M.L.; Wood, A.L.; Rao, P.S.C. *Environmental Science and Technology.* **1991**, *25,* 903-910.

20. Dai, D. Ph.D. Dissertation, University of Florida, Gainesville, Fl. **1997**.

21. Fingerling, G.; Herkorn, N.; Parlar, H. *Environmental Science and Technology.* **1996**, *30,* 2984-2992.

22. Paris, D.F., Lewis, D.L., Barnett, J.T. *Bulletin of Env. Contamination & Toxicology.* **1977**, *17,* 564-572.

23. Pearson, R.F., Swackhamer, D.L., Eisenreich, S.J., and Long, D.T. *Environmental Science and Technology.* **1997**, *31,* 3523-3529.

24. Augustijn, D.C.M. Ph.D. Dissertation. University of Florida, Gainesville, FL, **1993**.

25. Augustijn, D.C.M.; Dai, D.; Rao, P.S.C.; and Wood, A.L. In *Transport and Reactive Processes in Aquifers;* A.A. Balkema: Rotterdam. Bookfield. **1994**; pp 557-562.

26. Bouchard, D.C. *Contaminant Hdrology.* **1998**, *34,* 107-120.

27. Wauchope, R.D.; Buttler, T.M.; Hornsby, A.G. *Reviews of Environmental Contamination and Toxicology.* **1992**, *123,* 149.

28. Park, G.S. *J. Polym. Sci.* **1953**, *11,* 151.

Chapter 4

In-Place Regeneration of Granular Activated Carbon Using Fenton's Reagents

C. L. De Las Casas[1], K. G. Bishop[1], L. M. Bercik[2], M. Johnson[1],
M. Potzler[1], W. P. Ela[1,*], A. E. Sáez[1], S. G. Huling[3],
and R. G. Arnold[1,*]

[1]Chemical and Environmental Engineering, The University of Arizona,
Tucson, AZ 85721
[2]Tetra Tech FW, Inc., 1230 Columbia Street, San Diego, CA 92101
[3]Office of Research and Development, National Risk Management Research
Laboratory, U.S. Environmental Protection Agency, Ada, OK 74820

The feasibility of using Fenton's reagents for in-place
recovery of spent granular activated carbon (GAC) was
investigated. Fenton's reagents were cycled through spent
GAC to degrade sorbed chlorinated hydrocarbons. Little
carbon adsorption capacity was lost in the process. Seven
chlorinated compounds were tested to determine compound-
specific effectiveness for GAC regeneration. The contaminant
with the weakest adsorption characteristics, methylene
chloride, was 89% lost from the carbon surface during a 14-
hour regeneration period. Results suggest that intraparticle
mass transport limits carbon recovery kinetics, as opposed to
the rate of oxidation of the target contaminants. Fenton-
dependent recovery of GAC was also evaluated at a field site
at which GAC was used to separate tetrachloroethylene (PCE)
and trichloroethylene (TCE) from contaminated soil vapors. In
the field, up to 95% of the sorbed TCE was removed from

GAC during regeneration periods of 50-60 hours. Recovery of PCE was significantly slower. Although the process was not proven to be cost effective relative to thermal regeneration or carbon replacement, straightforward design and operational changes can lower process costs significantly.

Introduction

Ten of the 25 most frequently detected hazardous contaminants at National Priority List sites are chlorinated volatile organic compounds (VOCs) Trichloroethylene (TCE) and tetrachloroethylene (PCE) are among the top three (*1*). Both are moderately soluble and semi-volatile. Consequently, they are commonly recovered from contaminated groundwater or soils using pump-and-treat, air stripping, and soil vapor extraction (SVE) methods. Granular activated carbon (GAC) adsorption is commonly used to separate VOCs from liquid and gas streams derived from those recovery techniques. After the carbon is loaded to capacity, it must be regenerated or replaced. Advanced oxidation processes (AOPs) are also widely employed for the remediation of sediment and groundwater contaminated with VOCs (*2-4*). AOPs can destroy contaminants in place but are relatively expensive for treatment of low-concentration pollutants. Fenton's reaction is an AOP process in which reaction of hydrogen peroxide (H_2O_2) with iron (Fe) generates hydroxyl radicals (OH) that can react with a wide variety of VOCs (*5-6*). Fenton-dependent processes can mineralize even heavily halogenated targets such as PCE and TCE (*7*).

The present work explored the feasibility of using Fenton's reaction for in-place chemical regeneration of spent GAC. Rate limitations to Fenton-dependent regeneration of GAC loaded with halogenated VOCs were investigated in a series of bench-scale experiments. Process feasibility was also examined at an Arizona state Superfund site where carbon was used to separate chlorinated VOCs and volatile hydrocarbons from a contaminated gas stream. Carbon recovery kinetics at bench and field scales were compared.

The field site selected was the Park-Euclid (Tucson, Arizona) state Superfund site, where the primary vadose zone contaminants are PCE, TCE, dichloroethene (DCE) isomers, and volatile components of diesel fuel. A SVE system with GAC treatment of the off-gas was installed at the site as an interim remediation scheme while the state Remedial Investigation at the Park-Euclid site was underway.

There are four distinct zones of contamination–the upper vadose zone, perched aquifer, lower vadose zone, and regional aquifer (Figure 1). Both the upper and lower vadose zones contain dry-cleaning-related contaminants (i.e., PCE, TCE, and DCE isomers). The Park-Euclid SVE system draws gases from the upper vadose zone. Chlorinated solvents (primarily PCE) are present in both free product (diesel fuel atop the perched aquifer, Figure 1) and perched groundwater. The regional aquifer begins at about 200 ft below land surface. Diesel fuel has not contaminated the regional aquifer, however, a PCE plume with concentrations from 1-100 ppb extends more than 1,300 feet north-northeast from the origin of contamination, a well serving a former on-site dry cleaner (8). TCE and degradation by-products such as cis-1,2-dichloroethene are also present in a plume. Although there are no production wells in the immediate vicinity, the regional aquifer is relied upon to provide potable water within the Tucson basin. Maximum aqueous-phase contaminant concentrations in the regional aquifer are on the order of 10 and 100 ppb for TCE and PCE, respectively (8).

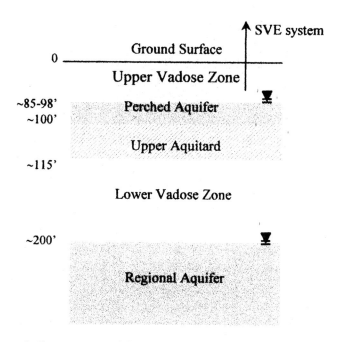

Figure 1. Cross-section of the Park-Euclid Arizona state Superfund site. PCE and TCE contamination is observed in the Regional and Perched Aquifers and in both the lower and upper vadose zones. SVE gases are extracted from the upper vadose zone.

Contaminant Removal/Destruction Mechanism

As shown in Figure 2, removal of adsorbates from GAC may be kinetically limited by any of four distinct steps: desorption from the GAC surface (labeled "**1**"); pore and surface diffusion (labeled "**2**"); film transport (labeled "**3**"), in which the film thickness (δ) is inversely related to the local bulk velocity; and removal of reactant from the bulk aqueous phase (labeled "**4**") due to reaction and advective transport. In most cases, intraparticle effects (pore diffusion, **2**) and/or desorption (**1**) control the observed rate of transfer from the particle (**1**) to the bulk aqueous phase (**4**) (*9*).

Figure 2 illustrates the possible sources of rate limitation experienced by a desorbing species. Partitioning between the solid and liquid within a pore is governed by the surface desorption rate or equilibrium condition (**1**). Once the contaminant is in the liquid within a pore, intraparticle transport (**2**) is largely governed by molecular diffusion. At the surface of the particle, transport into the bulk aqueous phase can be limited by molecular diffusion across a mass transfer boundary layer. Removal from the bulk aqueous phase relies on a mixture of convection and (in this case) reaction.

Experimental data presented in this work suggests that intraparticle mass transfer controls the rate of contaminant removal from the solid to the bulk

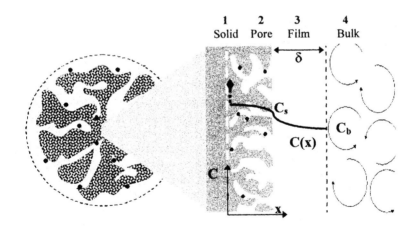

Figure 2. Potential sources of rate limitation for desorption of adsorbate from the GAC (solid) to the bulk fluid 1. Desorption from solid to liquid phase. 2. Diffusive transport within the pores (pore or surface diffusion). 3. Diffusive transport through a quiescent film surrounding the particle. 4. Convective transport or reaction in bulk fluid

liquid. Reaction of the target compound with hydroxyl radicals is assumed to occur mainly in the bulk aqueous phase. This physical model represents a hypothesis that can be tested using experimental carbon recovery data.

In the general case, it is possible that carbon recovery kinetics are limited by the rate of reaction of target contaminants with Fenton-dependent radicals, by rate of desorption of contaminants from the particle surface or by a combination of pore and surface diffusion. The physical situation hypothesized merely provides a starting point for model development and data analysis.

Materials and Methods

All chemicals were reagent-grade or better and were used as obtained. Granular activated carbon (URV-MOD 1, Calgon Corporation, Pittsburgh, PA) was used in all trials. This experimental-type carbon was selected because of its relatively high iron and low manganese contents. The GAC was steam-activated to minimize reactivity with H_2O_2 (10). It is a bituminous coal, 8 x 30 mesh (effective size 0.6-2.4 mm), with a specific surface area of 1290 m^2/g and pore volume of 0.64 mL/g (10). Carbon was rinsed with de-ionized water, dried at 103 °C, weighed, and then hydrated by soaking in deionized water for at least a day prior to use.

Pre-loading of carbon for bench-scale experiments was accomplished by placing 12-16 g of GAC into 1 L of water containing the experiment-specific target contaminant. Normally, the water was saturated with the target at room temperature prior to the introduction of carbon. The 1-L batch reactors with negligible headspace were then sealed and tumbled for approximately 60 hours at room temperature for attainment of equilibrium between the dissolved and adsorbed chemical. Final (measured) adsorbed concentrations are provided in Table I. In a subset of the field trials (see below), carbon-packed beds were loaded using the off-gases from the SVE system already in place. This was more representative of potential field operations.

Two types of laboratory experiments were conducted. Reaction kinetics were first studied by circulating Fenton's reagents through a pre-loaded, GAC-fluidized bed. Column recovery was monitored by periodically withdrawing carbon samples and extracting residual contaminants for analysis. Preliminary experiments of this type were conducted at room temperature. When it became apparent that room temperature was subject to significant uncontrolled changes, the temperature was controlled at 32°C in subsequent experiments of similar nature. In the second type of experiment, desorption rates were measured by circulating clean (contaminant-free) water through the column. That is, elution water did not contain Fenton's reagents. Water was passed through the pre

Table I. Chemical Properties of the Organic Compounds Studied

| Name | Log K_{ow} [a] | $k_{M,OH}$ $(M^{-1} s^{-1})$ [b] | Diffusivity (cm^2/s) [c] | Adsorbed concentration (mg/g) [d] | Freundlich Parameters [e] | |
					K (mg/g) (L/mg)$^{1/n}$	1/n
Methylene Chloride (MC)	1.15	9.00E+07	1.21E-05	275	0.07	1.06
1,2-DCA	1.47	7.90E+08	1.01E-05	146	0.04	1.33
1,1,1-TCA	2.48	1.00E+08	9.24E-06	20	0.65	0.87
Chloroform (CF)	1.93	5.00E+07	1.04E-05	188	1.48	0.77
Carbon Tetrachloride (CT)	2.73	2.00E+06	9.27E-06	N/A	12.30	0.59
TCE	2.42	2.90E+09	9.45E-06	103	5.82	0.70
PCE	2.88	2.00E+09	8.54E-06	11	45.66	0.56

NOTE: $k_{M,OH}$ is the second-order rate constant for the reaction of hydroxyl radical with the target organic compound.

SOURCE: [a]Reference 13; [b]Reference 12, except carbon tetrachloride (14); [c]calculated from Wilke-Chang equation (15); [d]From analysis of initial carbon concentration for carbon recovery experiments (data at 32°C). [e]From isotherm data obtained in this lab at 32°C (not shown).

loaded columns at rates designed to minimize aqueous-phase contaminant concentrations at the column exit, so that rates of mass transport from the carbon surface to bulk solution were not affected by the bulk liquid-phase concentration. In this manner, contaminant transport kinetics from surface to bulk could be established directly. Desorption kinetics were established by periodically measuring both residual contaminants on carbon samples and the contaminant concentration in the liquid exiting the column. All of these experiments (Fenton's reagents not present) were conducted at 32°C.

In the initial bench-scale column experiments, VOC-loaded carbon was packed into a Chromaflex® borosilicate glass chromatography column, I.D.=2.5 cm, L=15 cm, V=85 mL (Ace Glass, Inc., Louisville, KY). Fenton's reagents were prepared on the day of use. During regeneration, a 10 mM Fe (ferric sulfate) solution was recirculated in up-flow mode through the column at a rate sufficient to expand the carbon bed by approximately 50%. To initiate recovery, 0.2 M H_2O_2 was added to the recirculating fluid. At 10-60 minute intervals, sufficient H_2O_2 was added to restore the original concentration. This generally produced an H_2O_2 concentration that differed from the original concentration by less than 50% (data not shown). Carbon samples were periodically withdrawn from the top and bottom of the column for extraction and analysis of the target compound. Extraction periods were 12 hours in ethyl acetate on a shaker table. Extracts were analyzed using GC-ECD using methods described below. Carbon samples were then dried at 103°C and weighed. Data are reported as the mean results using samples from the top and bottom (one each) of the fluidized bed. Aqueous-phase samples were taken from the recirculation reservoir for analysis of the target compound, reaction by-products and residual hydrogen peroxide.

The clean-water recovery experiments were carried out to determine if carbon recovery is a transport-limited process, even when Fenton's reagents are present. The protocol used paralleled that of the recovery experiments with iron and H_2O_2 in solution, but the reagents were omitted. Temperature was again maintained at 32 ± 2°C, but only MC, CF and TCE were used in this work. Bulk concentrations of contaminants were maintained far below equilibrium levels by continuously eluting the target compounds with clean water. This maximized the rate of mass transport out of the carbon into the bulk aqueous phase by maintaining the bulk contaminant concentration near zero.

The SVE system at the field site was reconfigured to provide a side stream of SVE gases (containing mainly TCE and PCE) to the GAC column in order to load the carbon. The carbon was packed into a borosilicate glass chromatography column, I.D.=5 cm, L=30 cm, V=600 mL. The gas flow rate passing through the column was 4 cfm. Column effluent was returned to the SVE system. The carbon was loaded for approximately 72 hours. After loading, the carbon was regenerated in place via Fenton's reaction, reloaded with contaminant, and re-regenerated to study process feasibility. During regeneration, solid and aqueous-phase samples were withdrawn and extracted as in the bench-scale experiments. Initially, hydrogen peroxide was added to the 7 L-reservoir to maintain a near-constant concentration (0.2 M) throughout the regeneration period. Less frequent pulse additions of H_2O_2 were later used as a strategy to reduce H_2O_2 utilization during carbon recovery.

Analytical

Halogenated organic contaminants were analyzed by GC-ECD using a modified version of EPA Method 551.1. Twenty-μL samples were extracted in 1 mL of heptane. One μL of the resulting solution was then injected into a Hewlett Packard 5890 Gas Chromatograph (Palo Alto, CA) equipped with a DB-624 fused silica capillary column (J & W Scientific, Fulsom, CA; 0.53 mm ID, 30 m in length) and electron capture detector (ECD). Nitrogen and helium were used as the make-up and carrier gases. The gas flow rate was 26 mL/min. The temperatures of the oven, detector and inlet were 100°C, 275°C, and 150°C, respectively. Run times were from 5 to 10 minutes. A chlorinated compound (e.g., carbon tetrachloride) was added to each sample as an internal standard, and a check standard was run every tenth sample. Analyses were discarded if the deviation among check standards was larger than 5%.

Hydrogen peroxide was analyzed using a peroxytitanic acid colorimetric method (11) that was modified as follows. The sample (50 μL) and 50 μL of titanium sulfate (Pfaltz and Bauer, Inc., Waterbury, CT) solution were added to 4.9 mL of deionized water. Titanium sulfate was provided in stoichiometric excess to react with H_2O_2 leading to color development and quenching the Fenton reaction. After 1 hour, color development was measured at a wavelength of 407 nm using a Hitachi U-2000 doubled-beamed spectrophotometer (Hitachi Corporation, Schaumburg, IL). Samples were diluted with deionized water as necessary to fall within the range of the standards.

A Hach One pH/LSE meter (Hach Company, Loveland, CO) was used to monitor the pH of the solution containing Fenton's reagents. The meter was calibrated using standard pH calibrating buffers (pH 2 and 4) from VWR (Aurora, CO).

Results and Discussion

Bench-Scale Experiments

Contaminants were selected for Fenton-driven regeneration experiments to yield a range of compound hydrophobicities and reactivities with ·OH (Table I). Carbon recovery experiments were run at 24 ± 2°C for seven chlorinated VOCs (methylene chloride (MC), 1,2-dichloroethane (DCA), chloroform (CF), 1,1,1-trichloroethane (TCA), carbon tetrachloride (CT), TCE and PCE). Recovery data for MC, CF and TCE are provided in Figure 3. MC recovery was faster than CF, which was faster than the loss of TCE from the column reactors. The transformed data are shown on a semi-log plot in Figure 4. After an initial period of 1-3 hours, decay in the sorbed concentration conformed to first-order kinetics. TCE removal from GAC was only 50% complete after 14 hours. This was unexpected since the second-order rate constant ($k_{M,OH}$, Table I) for the reaction of hydroxyl radical with TCE (2.9E+09 $M^{-1}s^{-1}$) is near the molecular-collision diffusion limit (12) and is the highest among the three compounds tested. Lack of dependence of recovery kinetics on reaction rate with $\cdot OH$ suggests that the kinetics of Fenton-driven recovery of GAC is controlled by mass transport, as opposed to the rates of hydroxyl radical generation and radical reaction with contaminant targets.

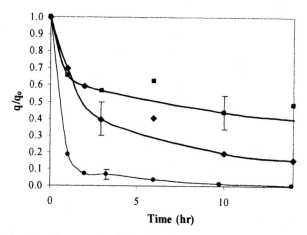

Figure 3. Fractional removal of adsorbate from GAC for MC (●), CF (◆) and TCE (■). Fractional q/q_o represents the mass of contaminant remaining in the carbon. The regenerant solution contained 10 mM iron, pH = 2.0, and 0.15 M H_2O_2 average concentration throughout the experiment. Temperature was controlled in the reservoir at 32 °C. The lines represent a smoothed fit to the data. Average error bars are indicated for each curve.

The shapes of the curves in Figure 3 are consistent with an intraparticle diffusion limitation: at short times, target concentrations in most of the GAC pores were nearly uniform, until a concentration profile developed along particle radii.

With time, the concentration gradient penetrated further into the particle increasing the apparent diffusion length and leading to slower, nearly first-order contaminant removal kinetics. The apparent first-order rate constants derived from the latter portion of each experiment are summarized in Table II.

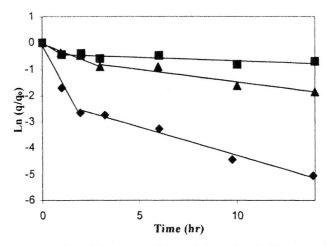

*Figure 4. Semi-log plot of the data in Figure 3 for MC (◆) CF (▲) and TCE (■).
The slopes from the long-time straight lines are used to calculate the observed
rates (k_{obs}).*

Experimental results (see Figure 6 and its discussion below) suggest that
mass transfer rates might provide the slow step in GAC regeneration for the
conditions employed in this work. If intraparticle diffusion is the controlling
transport mechanism, it would be expected that the concentration of target
compound in the liquid outside the particles to be relatively low due to its
consumption by Fenton's reaction. In that case, the flux of target lost from the
particle surface (J) should follow a relation like

$$J = k_m C_p \tag{1}$$

where k_m, is a mass transfer coefficient that represents intraparticle diffusion and
C_p is an average concentration of the target compound in the liquid that fills the
pores of the particles. If adsorption and desorption rates are fast, C_p would be at
equilibrium with the solid. Using the Freundlich isotherm, yields the following
expression:

$$C_p = \left(\frac{q}{K} \right)^n \tag{2}$$

To estimate the temporal variation of the target concentration in the solid,
consider the following mass balance of target:

Table II. Summary Table for Rates Observed (k_{obs}) at 24°C and 32°C and Cost Estimates at the Bench and Field Scales

Compound	K_{obs} (hr^{-1}) Bench-scale[a] $T=24$ °C	Bench-scale[b] $T=32$ °C	Bench-scale Desorption[c] $T=32$ °C	Field scale[d]	Cost ($/kg GAC)[e] Bench-Scale $T=32$ °C	Field-scale
MC	0.17	0.25	0.19	0.12	4.6	6.63 (100%) 3.57 (97%)
1,2-DCA	0.10	0.24	N/A	N/A	N/A	N/A
l,l,l-TGA	0.075	0.058	N/A	N/A	N/A	N/A
CF	0.066	0.11	0.11	0.059	3.50 (93%)	6.63 (93%)
CT	0.052, 0.060	N/A	N/A	N/A	N/A	N/A
TCE	0.045	0.027, 0.060	0.043	0.015 0.033 0.068	5.14 (52%)	2.55(73%) 4.53(82%) 6.54(95%)
PCE	0.037, 0.062	0.042	N/A	0.011	2.63(35%)	6.54(50%)

NOTE: [a]Bench-scale regeneration column experiments at 24°C (Figure 5), and [b]T=32°C (Figure 5). [c]Bench-scale desorption experiments (clean water, no Fenton reagents as eluant – Figure 6). [d]Field-scale regeneration column experiments. [e]Cost is based on hydrogen peroxide consumption for spent GAC regeneration. The number in parenthesis indicates the percentage GAC recovery for each trial.

$$M \frac{dq}{dt} = -JA_s \tag{3}$$

where M and A, are the total mass and total external surface area of GAC, respectively. It is assumed here that the target concentration in the solid is uniform, which would only be true at the start of the reaction cycle. It is also assumed that accumulation of target in the pore liquid is negligible compared to accumulation is the solid surface.

The preceding equation would have to be solved with the initial condition

$$q = q_0, \text{at } t = 0 \tag{4}$$

Using equations (1) and (2), equation (3) can be written as

$$M \frac{dq}{dt} = -A_s k_m \left(\frac{q}{K} \right)^n$$ (5)

This equation implies that the initial rate of decrease of the target concentration in the solid, dq/dt at $t = 0$, should be directly proportional to $1/K^n$ for all target compounds, providing that the initial concentration, q_0, is the same, and assuming that k_m would not vary appreciably for the various targets.

The preceding analysis led to the plot of the observed rate of target removal (k_{obs}) vs. $1/K^n$ for all target compounds. The results are shown in Figure 5. This analysis is probably too simplistic to try to establish a quantitative correlation for the time-dependent rate of target removal, but the fact that all compounds follow the expected trend (a higher $1/K^n$ implies a faster rate of removal) is consistent with the hypothesis that intraparticle mass transfer controls the process.

Results from clean water (no Fenton's reagents present) and Fenton-driven recovery experiments are compared in Figure 6. It is apparent that carbon recovery trajectories for the Fenton-driven and no-reaction (clean-water elution) cases matched very well for MG and TCE. For those contaminants, it seems evident that reaction did not limit recovery kinetics. For CF, however, degradation in the presence of Fenton's reagents was actually slower than in the clear water circulation experiments. This occurred because the bulk aqueous-phase concentration was held near zero in the latter experiments by continuously feeding clean influent to the column reactor. Since the reaction of ·OH with CF is relatively slow (Table I), CF accumulated to some extent in the bulk liquid when Fenton's reaction was relied upon to eliminate CF from the recirculated fluid. This diminished the driving force for CF transport from the GAC and protracted the recovery period. This interpretation is supported by measurements of liquid-phase CF concentrations during GAC recovery using Fenton's reagents (not shown). It is apparent that recovery kinetics were limited, at least in part, by the bulk liquid phase reaction of CF with ·OH.

Experiments Using a Field Column and Regeneration System

Equipment Testing–Methylene Chloride and Chloroform Recovery Tests

Initial field regeneration trials were carried out using a larger column (I.D. =5 cm, L=30 cm, V=600 mL, residence time =2 s) containing 100 g URV-MOD 1 GAC that was pre-loaded with MC or CF (under lab conditions). Contaminant

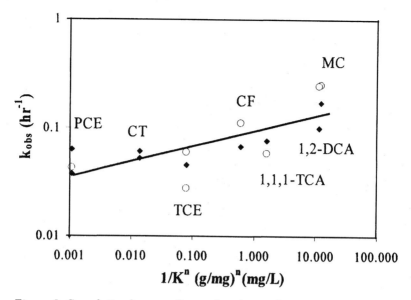

Figure 5. Correlation between first-order observed rate constant (k_{obs}) and compound-specific $(1/K)^n$. Rates for a 14-hour regeneration period with 10 mM iron, 0.1-0.15 M H_2O_2, pH = 2.0. Experiments were conducted at room temperature (24 °C) – closed symbols and 32 °C – open symbols. The line is a linear fit of the log-transformed data.

selection was based on hydrophobicity and reactivity with hydroxyl radicals (Table I).

Regenerant solution was recirculated continuously during each 30-hour experiment. Hydrogen peroxide was added at 1-hour intervals during hours 0-6 and 23-28 (Figure 7). At each point of addition, 150 mL of a 50% H_2O_2 stock solution (Fischer Scientific, Fairlawn, NJ) was added to the 7 L recirculation reservoir.

Carbon was periodically withdrawn from the top and bottom of the reactor for extraction and measurement of residual contaminants. Methylene chloride was essentially gone (full recovery) after 6-7 hours of operation. After 30 hours, just 6% of the original CF loading (125 mg CF/g carbon) remained on the GAC. The cost of recovery ranged from $2.5/kg to $6.6/kg GAC treated for the target contaminants studied. When multiple contaminants are simultaneously adsorbed to GAC, costs will be determined by the compound whose recovery is the slowest. As discussed later, little was done in these experiments to limit the non-

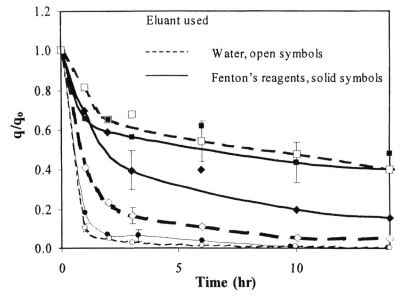

Figure 6. Comparison between recovery rates using eluant solutions with and without Fenton's reagents. When eluant consisted of water, bulk aqueous phase contaminants were near zero. Target compounds were MG (●), CF (◆), and TCE (■). Regenerant solutions contained 10 mM iron, 0.15 M average H_2O_2, pH = 2.0, T = 32 °C. Concentrations are normalized by the initial contaminant concentration on the carbon (q/q_o). Data points represent the average value between carbon extractions of top and bottom. An average error bar is indicated for one data point for each curve.

productive consumption of H_2O_2. That is, neither the configuration of the recovery system nor the schedule of H_2O_2 additions was designed to reduce H_2O_2 consumption/radical production that did not result in MG or CF destruction. Further discussion of this point is provided below.

Sequential Adsorption/Regeneration Experiments

The feasibility of carbon regeneration depends on both the acceptability of regeneration costs (primarily H_2O_2 consumption) and maintenance of carbon adsorption capacity through multiple degradation steps. Here, carbon adsorption

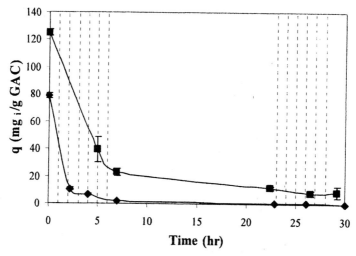

Figure 7. Carbon regeneration for (♦) MC and (■) CF in the field experiments. Units in the y-axis are "mg i/g GAC" for compound 'i." Degradation to below detection limit and 93% for MC and CF, respectively, was achieved after a 30-hour regeneration period. Reservoir concentrations: 10 mM iron, 0.15 M average H_2O_2, pH = 2.0. Dotted lines indicate times of 150 mL hydrogen peroxide additions. Error bars indicate the difference between top and bottom of the carbon extraction samples from the reactor.

capacity was tested before and after each of three surface regenerations. Carbon was loaded with 100-110 mg TCE/g GAC in a batch reactor in the lab, then transferred to the field site for regeneration in the field column. GAC (100 g, dry weight) was suspended in 1 L of pure water that was pre-saturated with TCE at room temperature (initial TCE concentration≈1100 mg/L). After 3 days, the distribution of TCE between carbon and liquid was near equilibrium with more than 99% of the contaminant on the carbon surface. The process was repeated twice using the same carbon sample to determine whether TCE adsorption was adversely affected by Fenton-driven regeneration. During regeneration periods, 0.7±0.2 g carbon samples were periodically withdrawn from the top and bottom of the column and extracted in ethyl acetate for determination of residual TCE. The regenerant solution containing 10 mM total Fe (pH 2) was recirculated at a rate that produced 50% GAC bed expansion. To initiate regeneration, 150 mL of 50% H_2O_2 was added to the 7 L regenerant volume to produce an initial H_2O_2

concentration of 0.38 M. Thereafter, the schedule of H_2O_2 additions was as indicated in Figure 8. At each point, an additional 150 mL of the 50% H_2O_2 stock solution was added to the regenerant reservoir.

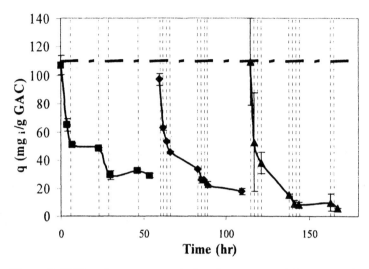

Figure 8. TCE carbon recovery during three sequential regeneration phases. Vertical dotted lines indicate points of hydrogen peroxide addition. The horizontal dashed line represents the TCE load (107 mg/g GAC) at the start of the first carbon recovery procedure. TCE degradation of 73% (■), 82% (◆) and 95% (▲) was obtained for the three consecutive regeneration cycles. Error bars indicate the difference between the carbon extraction values from top and bottom of the column.

In each phase of the experiment, carbon recovery was initially fast with 50% TCE loss from the carbon surface in 4 hours or less. Thereafter, recovery was much slower so that final (60-hour) TCE recoveries were 73, 82 and 95% during the sequential regenerations. Improvement in the later regeneration cycles was probably a consequence of more frequent H_2O_2 addition rather than a treatment-derived change in the physical characteristics of the URV-MOD1 carbon. Most important was the maintenance of TCE adsorption capacity over the 180-hour experiment (Figure 8). This finding is supported by previous investigations involving *N*-nitrosodimethylamine (2) and methyl *tert*-butyl ether (16) adsorption/regeneration on GAC.

TCE was also measured periodically in the regenerant reservoir to gage the adequacy of the H_2O_2 addition schedule. Comparison of regenerant TCE concentrations with (calculated) aqueous-phase concentrations in equilibrium with residual sorbed TCE concentrations (Figure 9) suggests that less frequent H_2O_2 could have produced similar recovery kinetics while reducing H_2O_2 consumption.

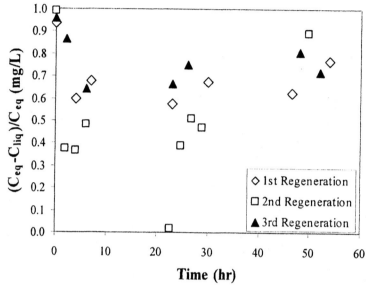

Figure 9. Ratio of ΔC_1 ($C_{eq} - G_{liq}$) to C_{eq}, where C_{eq} is the aqueous-phase TCE concentration in equilibrium with the residual adsorbed concentration (q) and C_{liq} is the measured, liquid phase concentration. Results for three consecutive regeneration periods are superimposed. Equilibrium concentrations were calculated using measurements of residual adsorbed TCE (Figure 8) and TCE isotherm parameters (Table I). Note: t=0 marks the beginning of each recovery cycle.

Were liquid levels to approach equilibrium with residual adsorbed TCE between peroxide additions, then reactor performance could be improved significantly by increasing the frequency of H_2O_2 additions. Conversely, if aqueous-phase concentrations remained low relative to equilibrium levels calculated on the basis of adsorbed mass, then the period of H_2O_2 addition could be extended to lower operational costs. The data suggest that dissolved TCE was quickly destroyed following each H_2O_2 addition to the regenerant. However, aqueous-phase TCE concentrations also recovered quickly after H_2O_2 was exhausted. There was apparently little to gain by decreasing the frequency of H_2O_2 addition in the experiment.

Temperature was measured in the regenerant solution during the experiment. Overall, temperature increased from ambient (~30°C) to 55-60°C during the 180-hour procedure. Temperature decreased slowly following H_2O_2 exhaustion (1-2°C/hr). Greater temperature increases might be expected in larger reactors although more judicious application of H_2O_2 or reduction in regenerant iron levels would tend to mitigate temperature rise. Because TCE mass transport and reaction kinetics are favorably affected by higher temperature, the exothermic decomposition of H_2O_2 via reaction with iron could, if handled carefully, increase carbon recovery rates and lower overall costs for carbon surface regeneration.

Loading Carbon with SVE Gases

Vadose zone gases from the soil vapor extraction system at the Park-Euclid (Arizona) state Superfund site containing primarily PCE, TCE and light diesel components as contaminants were used to load URV-MOD 1 GAC in a final set of field experiments.

The GAC was loaded for approximately 72 hours using a 4-cfm SVE sidestream. Effluent gases were pumped back into the system of extraction wells. To determine the carbon loading, GAC samples were taken from the top and bottom of the column, extracted in ethyl acetate and analyzed with GC-ECD. Initial, 6-hour regeneration trials produced 80% reduction in the adsorbed TCE concentrations but only a 30% loss of adsorbed PCE (Figure 10). Again, degradation was initiated by adding 150 mL of the 50% H_2O_2 stock to the regenerant solution (total volume 7 L). Subsequently, 50 mL of the stock H_2O_2 solution was added every 15-30 minutes to replenish the initial H_2O_2 concentrations throughout the regeneration period. This procedure led to excessive H_2O_2 utilization and attendant cost.

A total of 1.1 L of 50% H_2O_2 was consumed to destroy 7.0 g of PCE and 4.0 g TCE. Pulsed addition of H_2O_2 with intervening periods in which H_2O_2 was exhausted (for contaminant transport to the bulk regenerant phase) might have produced comparable recoveries using a fraction of the oxidant. Peroxide costs could also be lowered significantly by reducing the volume of regenerant in the system. In the presence of Fenton's reagents, aqueous-phase concentrations of contaminants were generally near zero. Because the rate of H_2O_2 consumption is independent of the contaminant concentration, however, H_2O_2 use continued during such periods without affecting contaminant transport out of the carbon particles. Under all circumstances, H_2O_2 consumption was proportional to the total regenerant volume, including the volume in the recirculation tank, where little or no contaminant was consumed under conditions of the field test.

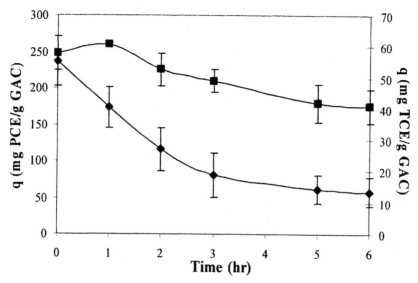

Figure 10. Carbon regeneration for SVE-loaded GAC. The two primary pollutants at the site are PCE and TCE. Overall, 30% and 80% degradation for PCE (■) and TCE (♦) was achieved during the 6-hour regeneration period. Reagent concentrations were 10 mM iron, 0.15 M H_2O_2 (average), at pH = 2.0. Error bars indicate the difference between the carbon samples from top and bottom of the column.

Economic Analysis

Sources of cost for an in-place GAC recovery system based on Fenton's mechanism would include reagent (H_2O_2) consumption, power requirements to circulate regenerant, replacement of carbon that is lost due to abrasion and chemical (treatment) inactivation, capital expenses associated with the regeneration system itself (pumps, pipes and tanks) and the additional capacity necessary for temporary column retirement for regeneration. Here, it is assumed that H_2O_2 costs dominate the overall cost of carbon recovery. The primary sinks for ·OH are H_2O_2 itself, due to concentration effects used, and, possibly (as experiments or regeneration cycles progress), free chloride ion. Thus, free radical concentrations (and H_2O_2 consumption rates necessary to maintain those levels) are essentially independent of the identity or concentration of the target compounds. This is not to say, however, that H_2O_2 costs are independent of contaminant identity. Compounds that desorb slowly from GAC require significantly greater time to achieve comparable degrees of recovery and, hence, greater recovery costs.

When multiple contaminants are present simultaneously, the compound most resistant to Fenton-driven recovery, in this case PCE, is likely to dominate recovery costs. Consequently, PCE degradation was used to estimate overall carbon regeneration costs. Other assumptions and economic or operational factors follow:

Unit cost of H_2O_2 (50% solution, 1.18 g/mL)	$0.341 (2) (transportation cost not included)
H_2O_2 utilization for carbon recovery	95-232 mL (bench-scale column) 1-2 L (field column)
Carbon in experimental columns	12-16 g (bench-scale column) 78-100 g (field column)
Carbon purchase cost (17)	$1.54-2.64/kg virgin coal carbon $1.10- .72/kg regenerated carbon
Carbon change out/disposal	$0.66/kg (18)

Costs for carbon recovery/replacement alternatives are compared in Table III. The comparison is admittedly crude. The mechanism, rate limitation and kinetics of PCE recovery on GAC are poorly known, certainly not well enough to produce a reliable economic analysis. In-place oxidations of all other contaminants tested were significantly faster than that of PCE and, therefore, more economically attractive. Nevertheless, PCE was among the important contaminants in this application and should be considered when assessing the utility of the technology locally.

Methods for increasing the efficiency of H_2O_2 use have been discussed to some extent. The volume of the mixing reservoir allowed the Fenton reaction to take place and H_2O_2 to be consumed without oxidation of the target contaminants. Based on the dimensions of the pilot-scale column, and assuming a porosity of 0.5 of the GAC, the pore volume within the column is

Table III. Comparison of GAC Replacement *vs.* Regeneration Costs

Option	Cost/kg	Critical assumptions/parameters
1. Replacement	$3.30	$2.64/kg purchase cost
		$0.66/kg disposal of spent carbon
2. Thermal Regeneration	$2.64	$1.65/kg regeneration cost
		$0.66/kg transportation
		$0.33/kg carbon replacement
3. Fenton-based, in-place		Peroxide costs dominate (0.34/L)
recovery		PCE recovery dictates treatment time
a. Lab column	$2.69	95 mL H_2O_2/12 g GAC
b. Field column	$6.54	1.5 L H_2O_2/78 g GAC

approximately 0.3 L. The total volume of the reactor system was 7 L. This suggests that more than 95% of the H_2O_2 applied to this system was consumed outside of the column (i.e., 0.3 L/7 L). Although it was assumed that the Fenton reaction (in the experimental system) occurred predominantly in the mixing reservoir, limiting the H_2O_2 reaction outside the mixing reservoir would economize H_2O_2. It may be possible to immobilize iron on the carbon surface and run regenerations in the optimal pH range to avoid iron dissolution during recovery operations. The feasibility of such a scheme depends on selection of iron loadings that allow degradation reactions to proceed without blocking the carbon surface or interfering with contaminant access to carbon pores. Preliminary tests of Fenton-driven recovery in such iron-mounted systems are underway. Their potential advantage lies in localization of H_2O_2-consuming reactions and radical generation in the vicinity of the carbon surface. Bulk-aqueous-phase Fenton reactions can be minimized so that non-productive H_2O_2 consumption (that which destroys no contaminants) is essentially eliminated.

Data suggest that a single mechanism does not control the rate of regeneration for all contaminants. Poorly adsorbed compounds with relatively low reactivity with ·OH, like chloroform, can be limited by reaction in the bulk aqueous phase. Less soluble, more reactive compounds like TCE are limited by intraparticle transport. Consequently, optimal approaches to carbon recovery may be compound-dependent. Pulsed addition of H_2O_2 offers advantages over continuous maintenance of a target H_2O_2 concentration when mass transport governs the carbon recovery rate. Temperature management may be an essential issue inasmuch as the kinetics of physico-chemical processes of importance are temperature dependent. Although Fenton-based carbon regeneration was not proven to be cost effective in this application, cost efficiency can be improved

64

substantially by implementing design/operational changes discussed or perhaps, more dramatically, by directing radical-generating reactants to the carbon surface.

Notice

The U.S. Environmental Protection Agency, through its Office of Research and Development, funded and managed the research described here under CR 82960601-0 with the University of Arizona. It has not been subjected to Agency review and therefore does not necessarily reflect the views of the Agency, and no official endorsement should be inferred.

References

1. National Research Council. *Alternatives for Ground Water Cleanup;* National Academy Press: Washington, DC, 1994; p 315.
2. Kommineni, S.; Ela, W.; Arnold, R.; Huling, S.; Hester, B.; Betterton, E. *Environ. Eng Sci.* **2003**, *20,* 361-373.
3. Prousek, J. *Chem. Lisy.* **1995**, *89,* 11-21.
4. Zoh, K.; Stenstrom, M. *Wat. Res.* **2002**, *36,* 1331-1341.
5. Gallard, H.; De Laat, J. *Wat. Res.* **2000**, *34,* 3107-3116.
6. Huling, S.; Arnold, R.; Jones, P.; Sierka, R. *J. Environ. Eng.* **2000**, *126,* 348-353.
7. Teel, A.; Warberg, C.; Atkinson, D.; Watts, R. *Wat. Res.* **2001**, *35,* 977-984.
8. Miller Brooks Environmental Inc. *Final Draft Park-Euclid Remedial Inevestigation Report;* Arizona Department of Environmental Quality, Project No. 365-0011-01; Phoenix, AZ, 2004; p 707.
9. Crittenden, J.C.; Hand, D.W.; Arora, H.; Lykins, B.W. *J AWWA.* **1987**, *79,* 74-82.
10. Huling, S.G.; Jones, P.K. Submitted to Environ. Sci. Technol., **2005**.
11. *Colorimetric Determination of Nonmetals;* Boltz, D.; Holwell, J., Eds.; 2nd. Edition; Chemical Analysis 8; John Wiley and Sons, Inc.: New York, 1978; p 543.
12. Radiation Chemistry Data Center of the Notre Dame Radiation Laboratory, URL http://www.rcdc.nd.edu/ (May, 2003).
13. Swarzenbach, R.; Gschwend, P.; Imboden, D. *Environmental Organic Chemistry;* John Wiley & Sons, Inc.: New York, 1993; p 681.
14. Haag, W.R.; Yao, C.C.D. *Environ. Sci. Technol.* **1992**, *26,* 1005-1013.
15. Logan, B. *Environmental Transport Processes;* John Wiley & Sons, Inc.: NewYork, 1999; p 654.

16. Huling, S.; Jones P.; Ela, W.; Arnold, R. *Wat. Res.* **2005**, *39,* 2145-2153.
17. U.S. EPA. Wastewater Technology Fact Sheet. Granular Activated Carbon Absorption and Regeneration, U.S.EPA-Office of Water: Washington, DC, 2000; EPA 832-F-00-017.
18. Environmental and Remediation Equipment, URL http://www.soiltherm.com/GAC_Comparison/gac_comparison.html (August, 2005).

Chapter 5

Sorption and Hydrolysis of Environmental Pollutants by Organoclays

Zev Gerstl[1], Ludmila Groisman[2], Chaim Rav-Acha[2], and Uri Mingelgrin[1]

[1]Institute of Soil, Water and Environmental Sciences, The Volcani Center, ARO, P.O. Box 6, Bet Dagan 50250, Israel
[2]Research Laboratory of Water Quality, Ministry of Health, P.O. Box 8255, Tel-Aviv 61080, Israel

Organoclays are clays whose surfaces are rendered organophilic by the exchange of inorganic cations with various organic cations. The sorption of six compounds with a range of log K_{ow} values from 2.5 to 6 was studied on short- and long-chain organoclays. Compounds with low or medium hydrophobicities were more strongly sorbed on the short-chain organoclay, whereas the more hydrophobic compounds were better sorbed on the long-chain organoclay. It was found that both types of organoclays were able to remove organic pollutants from the effluent of a pesticide producing plant, but solute uptake by short-chain organoclays was strongly depressed by competition, while long-chain organoclays were only slightly affected by the presence of competing solutes in the industrial wastewater.

A bifunctional organoclay that is able to sorb organophosphate pesticides, as well as to catalyze their hydrolysis, has been prepared. The detoxifying capacity of this organoclay for methyl parathion and tetrachlorvinphos, was demonstrated.

The half-life for the hydrolysis of the investigated pesticides in the presence of the bifunctional organoclay is about 12 times less than for their spontaneous hydrolysis. The mechanism of the catalytic hydrolysis of methyl parathion by the bifunctional organoclay was studied by following the effect of replacing H_2O with D_2O, by replacing the primary amino headgroup of the organic cation in the bifunctional clay by a tertiary amino group and by a detailed mathematical analysis of the reaction kinetics. An isomer of MP was formed in the presence of the bifunctional organoclay, initially increasing in concentration and then disappearing, the effect of the isotope replacement was minimal and the tertiary amine substitution increased the rate of MP hydrolysis. Based on these findings we propose a mechanism in which hydrolysis of MP proceeds both via a direct route (specific base hydrolysis) and via the formation of the isomer which then undergoes specific base hydrolysis more rapidly than the parent MP. The relative importance of each pathway is a function of pH with the direct hydrolysis of MP predominant at higher pH values (pH>10) and the isomer pathway predominating at intermediate pH values (pH ~ 8-10).

Introduction

Organoclays are used in a number of varied applications, such as polymer–clay nanocomposites (*1-3*), absorbents of organic pollutants in ground water, coatings and paints (*4-8*). Organoclays are smectite clays whose surfaces are rendered organophilic by the exchange of inorganic cations with various organic cations, mostly quaternary ammonium compounds (QACs). This exchange leads to an increased sorption affinity for non-ionic organic contaminants caused by the alkyl chains of the QAC, which hydrophobize the clay surface.

Organoclays may be divided into two groups depending on the structure of the organic cation and the resulting mechanism of sorption (*9*). The first group, called sorptive organoclays, includes clays that contain short-chain quaternary ammonium ions. Sorption on this type of organoclay is characterized by Langmuir-type isotherms that are commonly associated with specific sorption sites. The second group of organoclays, called organophilic organoclays, is composed of clays that contain long-chain quaternary ammonium ions. Sorption by this group is characterized by linear isotherms over a wide range of solute concentrations.

It is generally believed that the short-chain sorptive organoclays sorb non-ionic organic compounds more effectively than the more bulky organic phase of the long-chain organoclays (LCOC) (*10-12*). This view is based on a series of studies carried out in large part with low molecular weight compounds that are not highly hydrophobic, in which it was found that the sorption capacity of short-chain organoclays was much higher than that of the long-chained organoclays (*11*). The results were explained by the different mechanisms of sorption by which the two classes of organoclays take up sorbates. However, the above assertion regarding the higher sorption capacity of the short-chained as compared to the long-chained organoclays was shown in the present study to be invalid for highly hydrophobic compounds.

Degradation of environmental pollutants is of particular interest because it eliminates the contaminants completely, whereas, by sorption, the chemical is simply moved from one compartment to another in a given system. in order to enhance the potential of organoclays for removing pollutants from aqueous environments, a novel organoclay, termed a bifunctional organoclay, was developed, in which a second functional group was introduced on one of the hydrocarbon chains of the quaternary ammonium cation. Specifically an aminoethyl group was chosen as a second functional group since free alkylamines attack organophosphate and other esters and catalyze their hydrolysis.

Herein, we briefly review findings of our research group on the sorptive properties of organoclays and the utility of the bifunctional organoclay as a catalyst.

Materials and Methods

Detailed descriptions of the materials and methods used can be found in our previously reported work (*13-15*). Below we will present only brief descriptions of the experimental approaches used in this work.

Materials

Cation Synthesis

Two long-chain organic cations of the structure,

$$CH_3$$
$$\overset{+}{|}$$
$$C_{10}H_{21}\text{-}N\text{-}(CH_2)_2\text{-}NX_2$$
$$|$$
$$CH_3$$

where X can be either H or CH_3, were synthesized for the present study.

Synthesis of the N-decyl-N,N-dimethyl-N-(2-aminoethyl) ammonium cation (DDMAEA; X=H) and of N-decyl-N, N-dimethyl-N-(2-N, N-dimethyl aminoethyl) ammonium (DDMAEDMA; X=CH_3) has been described previously (*14* and *15*, correspondingly). A short-chain bifunctional organoclay was obtained from *N, N, N*-trimethyl-*N*-(2-aminoethyl) ammonium (TMAEA) chloride (Aldrich, St. Louis, MO).

Organoclays

Organoclays were prepared by mixing the bentonite clay with solutions containing the QACs. All quaternary ammonium salts (Aldrich, St. Louis, MO) were of 98% purity or higher. The QAC salts used to prepare the organoclays included tetramethylammonium (TMA), octadecyltrimethylammonium (ODTMA), N-decyl-N, N, N-trimethyl ammonium (DTMA), and the long- and short- chained bifunctional cations (Table I).

Table I. The structure of the various QACs used for preparing organoclays.

$$R_1$$
$$|$$
$$R_3 - \overset{+}{N} - R_4$$
$$|$$
$$R_2$$

Organoclay	R_1	R_2	R_3	R_4
DTMA	CH_3	CH_3	$C_{10}H_{21}$	CH_3
ODTMA	CH_3	CH_3	$C_{18}H_{37}$	CH_3
DDMAEA	CH_3	CH_3	$C_{10}H_{21}$	$(CH_2)_2NH_2$
DDMAEDMA	CH_3	CH_3	$C_{10}H_{21}$	$(CH_2)_2N(CH_3)_2$
TMAEA	CH_3	CH_3	CH_3	$(CH_2)_2NH_2$
TMA	CH_3	CH_3	CH_3	CH_3

Solutes

Five pesticides and one polyaromatic hydrocarbon (PAH), covering a wide range of hydrophobicities, were selected for sorption studies. The names, structures, log K_{ow} (a measure of hydrophobicity) and aqueous solubilities of the selected compounds are shown in Table II. Methyl parathion (MP) and tetrachlorvinphos were chosen for the evaluation of the bifunctional organoclay as a catalyst (Table II).

Industrial Wastewater

Industrial wastewater (after primary settling) was collected from a pesticide manufacturing plant in Israel (*13*). A list of pesticides that were detected in the collected industrial effluent samples and their concentrations are shown in Table III.

Sorption Isotherms

Batch sorption measurements of the solutes listed in Table II from distilled water were performed in triplicate, leaving no headspace. The suspensions were shaken at 23°C for 24 h. The aqueous phase was then separated by centrifugation and the concentration of sorbate remaining was determined by gas chromatography (GC), as described below.

Adsorption isotherms in wastewater were conducted after the top oily layer of the wastewater was mechanically removed, the pH was adjusted to 11 and the effluents were purged with nitrogen for 24 h to remove volatile components. The pH was then reduced to 6.5, and effluent samples were spiked to the desired concentrations of the different pesticides, taking into account the concentrations of these pesticides originally present in the wastewater (Table III). The adsorption experiments were then performed as for distilled water.

Degradation Studies

The rate of MP hydrolysis in the presence of long-chain bifunctional organoclay was measured in parallel with 5 controls as follows: a) MP on Na-bentonite; b) MP on DTMA-clay; c) spontaneous hydrolysis of MP; d) MP hydrolysis in a solution of the DDMAEA cation monomer; and e) hydrolysis of MP in a suspension of the short-chain bifunctional organoclay. At various times the solid phase was separated from the liquid phase, and the effluent was analyzed by GC/MS (as described below) for residual MP and *p*-nitrophenol (p-NP) produced by hydrolysis. All experiments were run in triplicate.

Sorbed concentrations of MP and p-NP were similarly measured by extracting the clay with a methylene chloride:acetone mixture followed by GC/MS analysis. The degradation kinetics experiments with tetrachlorvinphos were performed in a similar manner.

Hydrolysis of MP by the Tertiary Amine Bifunctional DDMAEDMA-Clay

The hydrolysis of MP in the presence of the tertiary amine bifunctional organoclay (DDMAEDMA-clay) was carried out in parallel with the hydrolysis in the presence of the primary amine organoclay DDMAEA at pH 10.8 (where the amino headgroups of both cations are fully deprotonated), under exactly the same conditions.

Analytical Procedure

The concentrations of the chemicals in solution were determined according to USEPA Method 525.2 (Revision 2.0) for the determination of organic compounds in water by liquid–solid extraction and capillary GC/MS using a ThermoQuest Trace GC2000/MS Polaris instrument (ThermoQuest, Waltham, MA). A C18 extraction disk (47 mm) was used for solid phase extraction (SPE). The analytical column used was a DB-5MS column (J & W Scientific, Folsom, CA; 30 m x 0.25 mm; 0.25 μ thickness).

Calculations

The kinetic rate constants were calculated by integrating the kinetic equations and fitting to the experimental data, using the proper (normalized) initial conditions at t=0. For details see Rav-Acha et al. (*15*).

Results and Discussion

Adsorption Isotherms of Individual Compounds from Purified Water

Isotherms on the ODTMA-clay were linear for all the compounds studied, whereas most of the isotherms on the TMA-clay were non-linear, as can be seen from the sorption parameters presented in Table IV. Consequently, sorption data for the ODTMA-clay were fit to a linear isotherm, and that of the TMA-clay were fit to the Freundlich equation

Table II: Compounds Used in the Sorption Studies

Compound	Structure	$\log K_{ow}$	Aqueous Solubility $(mg\ L^{-1})$
Atrazine		2.47	33
Ametryn		3.07	200
Prometryn		3.34	33
Terbutryn		3.74	22
Pyrene		5.18	0.135
Permethrin		6.12	0.2
Methyl parathion		2.86	60
Tetrachlorvinphos		3.56	15

Table III. Pesticides in the Effluent of a Pesticide Producing Plant in Israel

Pesticide	Concentration (mgL⁻¹)	Pesticide	Concentration (mgL⁻¹)
Atrazine	2.6	Bromacil	0.2
Ametryn	9.7	Fluometuron	0.1
Prometryn	0.9	Linuron	0.3
Terbutryn	1.4	Propazine	<0.1
Trifluralin	0.4	Prometone	2.0
Terbutyazine	2.0	Alachlor	0.1

$$Q = K_f C_n \tag{1}$$

where Q ($\mu g/g$) is the solid phase concentration; K_f, the adsorption coefficient; n, an empirical constant; and C, the solute concentration (mg/l). When $n \approx 1$, as is the case with all isotherms on the ODTMA-clay, K_f is equivalent to the distribution coefficient, K_d. Normalized adsorption coefficients on the basis of organic carbon content, K_{oc}, were calculated (Table IV).

Table IV. Sorption Parameters for Uptake by Short-chain (TMA) and Long-chain (ODTMA) Organoclays

	log K_{ow}	TMA -clay K_f	n	ODTMA-clay K_d	K_{oc}
Atrazine	2.47	1738	0.903	40.0	276
Ametryn	3.05	25119	1.26	173	1193
Prometryn	3.34	1349	0.84	380	2621
Terbutryn	3.74	2483	1.08	415	2862
Pyrene	5.18	84938	1.05	393000	2710340
Permethrin	6.12	22131	0.68	992000	6841380

The least hydrophobic compounds, atrazine, ametryn, prometryn and terbutryn, are sorbed considerably stronger by the short-chained TMA-organoclay than by the long-chained ODTMA-organoclay. The most hydrophobic compounds (pyrene and permethrin) are sorbed more strongly on the ODTMA-organoclay. Smith and Galan (11) and Smith et al. (10) stated that short-chain organoclays are in general much better sorbents than long-chain organoclays. They showed, for example, that tetrachloromethane is adsorbed 10 times more on a short-chained benzyltrimethylammonium-bentonite than on a long-chained hexadecyl-trimethylammonium-bentonite. The present results show that the above statement may be correct for low molecular weight compounds of relatively low hydrophobicity. However, for highly hydrophobic compounds the opposite is true. These are better sorbed on the long-chain organoclays. Thus, as indicated in Table IV, both pyrene and permethrin are adsorbed by the ODTMA-clay considerably more than by the TMA-clay.

The greater sorption observed for the less hydrophobic pesticides on short-chained organoclays, such as the TMA-clay, may be explained as follows. Due to the absence of a long hydrophobic chain in the TMA-clay, the original surface of the negatively charged clay, as well as the positively charged N-ammonium counter cations, are more accessible to interaction with sorbate molecules in this clay than in the ODTMA-clay. The less hydrophobic pesticides possess polar groups that can strongly interact with the charged sites at the surface and possibly undergo specific interactions, such as H-bonding, thereby enhancing sorption on the TMA-clay beyond that observed on the ODTMA-clay where access to the charged surface sites is blocked by the bulky octadecyl chains. The highly hydrophobic pyrene and permethrin are less prone to interact with charged surface sites and hence will display a higher sorption in the presence of the bulky octadecyl chain where a larger hydrophobic surface area (or 3-D volume) is available for hydrophobic sorption (16), than in the absence of this chain. When the K_{oc} values for the various compounds on the long-chain organoclay (Table IV) are plotted against the hydrophobicity indicator, K_{ow}, (Figure 1) a linear relationship is obtained, indicating that sorption on long-chain organoclays is highly influenced by hydrophobicity.

Adsorption of Pesticides from Mixtures

The ability of the organoclays to remove pesticides from mixtures in tap water and from wastewater spiked with the 4 relatively less hydrophobic pesticides (Table IV) was studied and compared with the respective isotherms measured in single-solute solutions in distilled water. The results are summarized in Tables V and VI. One significant observation is that, relative to sorption from single-solute systems on the short-chain TMA-clay, sorption from pesticide mixtures in tap water was low and was even more strongly depressed in wastewater. Sorption of the pesticides on the long-chain ODTMA-clay, on the other hand, was not very different when it took place in purified water or from mixtures in tap water or wastewater (Table VI). Despite the greater depressing effect of wastewater on sorption on TMA-organoclay, this short-chained

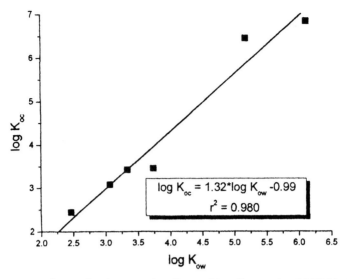

Figure 1. Relationship between log K_{ow} and log K_{oc} on the ODTMA-clay (based on Groisman et al. (13)).

organoclay still sorbed the lowest K_{ow} pesticides to a greater extent than did the long-chained ODTMA-organoclay.

Sorption on short-chain organoclays is likely to be a site-specific surface process, which is very sensitive to competition (*17*). In the case of sorption of the compounds from a mixture in tap water, the presence of the other pesticides in solution is, therefore, the most likely reason for the decrease in sorption. The competition in the case of wastewater is yet stronger since it stems not only from the accompanying pesticides but also, and perhaps mainly, from background materials (such as solvents), which are present in high concentrations. The effect of the competing sorbates on sorption by ODTMA-clay is limited, and the free sorption capacity of the long-chain organoclay remained largely unaffected by the presence of the extraneous compounds. This is so because the long alkyl chains provide numerous and non-specific hydrophobic sorption sites, or rather a large 3-D uptake volume, that reduce the importance of competition.

It is evident that short-chain organoclays sorb less hydrophobic non-ionic organic compounds better than do the long-chain organoclays; however, it seems that the long-chain organoclays are superior sorbents for highly hydrophobic compounds. Furthermore, sorption of low to moderately hydrophobic compounds by long-chain organoclays is only slightly affected by the presence

Table V. Comparison of Sorption of Triazine Herbicides on TMA-Clay from: Single-Solute System, Mixture in Tap Water and Mixture in Wastewater

	Single-solute		Mixture in tap water		Mixture in wastewater	
	K_f	n	K_f	n	K_f	n
Atrazine	1738	0.90	676	0.71	110	0.465
Ametryn	25119	1.26	603	0.39	372	0.608
Prometryn	1349	0.84	724	1.06	70.8	0.525
Terbutryn	2483	1.08	1259	0.95	251	0.666

Table VI. Comparison of Sorption of Triazine Herbicides on ODTMA-Clay from: Single-Solute System, Mixture in Tap Water and Mixture in Wastewater

	Single-solute		Mixture in tap water		Mixture in wastewater	
	k_d	r^2	k_d	r^2	k_d	r^2
Atrazine	40.0	0.981	22.5	0.958	14.7	0.803
Ametryn	173	0.985	153	0.980	121	0.981
Prometryn	380	0.985	230	0.981	176	0.993
Terbutryn	415	0.955	583	0.984	466	0.999

of competing extraneous sorbates and the removal of organic compounds, such as pesticides, from wastewater will be, therefore, less affected by competition with accompanying compounds and background materials when long-chain organoclays are used as sorbents than when using short-chain organoclays. The long-chain organoclays may thus be the preferred sorbent to remove pesticides from industrial wastewater.

Despite the high sorptive capacity of organoclays for nonionic organic pollutants, the sorbed pollutants may eventually undergo desorption or saturate the sorbing complex. Their definitive removal from the system requires they undergo some form of degradation. This was accomplished with the aid of the bifunctional clays developed by our research group, as discussed below: Sorption of Organophosphates on the Bifunctional Organoclay.

Sorption isotherms of MP and tetrachlorvinphos on the bifunctional DDMAEA-organoclay and on the non-bifunctional DTMA-organoclay at pH 9.0 were linear, and the sorption data are summarized in Table VII.

Table VII: Sorption of Methyl Parathion and Tetrachlorvinphos by the DTMA- and Bifunctional-Organoclays.

	Bifunctional organoclay		DTMA- organoclay	
	k_d	r^2	k_d	r^2
Methyl Parathion	1.40	0.982	1.85	0.978
Tetrachlorvinphos	1.50	0.998	1.83	0.988

Linear isotherms are expected for sorption of organic solutes on long-chain organoclays in which the bulk of the long organic chains may be viewed as a quasi-partitioning phase. Both MP and tetrachlorvinphos sorb slightly less strongly on the bifunctional- than on the DTMA-organoclay, which precludes the existence of a significant specific interaction between the sorbate molecule and the ethylamino group of the bifunctional organoclay. The sorption coefficient of MP on the short-chain bifunctional clay was 4.31 L/g, three times greater than on the long-chain bifunctional clay.

The Catalytic Effect of the Bifunctional Organoclay

The rate of MP hydrolysis in the presence of the long-chain bifunctional organoclay and of the 5 controls is shown in Figure 2. The catalytic effect of the bifunctional organoclay, relative to spontaneous hydrolysis is readily apparent. The half-life of MP in the presence of the bifunctional organoclay is around 5 days, while the half-life for the spontaneous hydrolysis of MP calculated from the data presented in Figure 2 is 40 days. The rate of MP hydrolysis on the long-chain bifunctional organoclay is also much higher than that of its hydrolysis in the presence of the conventional (non-bifunctional) DTMA organoclay. The advantage of the long-chain bifunctional organoclay over the regular (non-bifunctional) organoclay for detoxifying contaminated waters is, therefore, obvious.

The most significant observation in Figure 2 is that, whereas the catalytic effect of the long-chain bifunctional organoclay is very pronounced, both the non-bifunctional long-chain organoclay (DTMA) and the short-chain bifunctional organoclay (TMAEA) as well as the mineral clay displayed a retarding rather than a catalytic effect on the pesticide's hydrolysis. The stabilizing effect of these sorbents may be explained as follows. The major pathway for the spontaneous hydrolysis of organophosphates is specific base catalysis, in which the phosphate ester is attacked by a hydroxyl ion. It is possible that sorption of the organophosphate to these clays protects the ester from base-catalyzed hydrolysis due to the repulsion of the hydroxyl ions by the excess negative charge at the clay's surface.

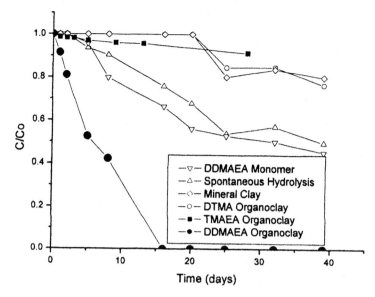

Figure 2. The rate of methyl parathion hydrolysis (pH 9.0) in the presence of the long-chain bifunctional (DDMAEA) organoclay and in 5 controls (based on Groisman et al. (14)).

The fact that only the long-chain bifunctional organoclay exhibits a catalytic effect on the hydrolysis of the organophosphate ester indicates that, for catalysis to take place, the presence of the relatively flexible long chain that can retain the substrate at a close proximity to the aminoethyl catalytic group is necessary. In contrast, even though sorption on the short-chain bifunctional organoclay is greater than on the long-chain one, steric considerations dictate that the proximity effect does not occur on this clay. This will be discussed below in more detail. The small catalytic effect of the cation monomer (Figure 2) may be attributed to the formation of micelles into which the pesticide molecules partitioned just as they did into the organic core of the long-chained bifunctional clay. The same general trend as that observed for the enhanced hydrolysis of methylparathion was found for tetrachlovinphos.

It was observed that the rate of MP disappearance was not matched by the rate of appearance of the main metabolite, *p*-nitrophenol (p-NP). This finding indicates that the mechanism by which the bifunctional organoclay hydrolyses MP is not a straight forward, one step reaction.

Identification of the Intermediate Product

In addition to the discrepancy between MP disappearance and p-NP formation observed soon after the beginning of the experiment, an extra peak in the chromatogram was observed in samples containing the bifunctional organoclay that first increased in area and then disappeared gradually towards the end of the reaction process. These findings lead us to the conclusion that the degradation of MP on the bifunctional organoclay proceeds through an intermediate. Based on the mass spectrum of this compound, we deduced an isomer, O, S-dimethyl-O-(4-nitrophenyl) thiophosphate, of the starting material. The appearance of an intermediate product and the fact that the kinetic curves are typical of consecutive first order reactions suggest the reaction scheme shown below where: MP, p-NP and I are methylparathion, p-nitrophenol and the intermediate, respectively, and k_1, k_2, k_3 are the first-order rate constants of the respective steps shown in Scheme 1.

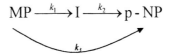

$$MP \xrightarrow{k_1} I \xrightarrow{k_2} p\text{-}NP$$

Scheme 1: A schematic description of the hydrolysis of MP catalyzed by the bifunctional-organoclay (DDMAEA).

Kinetic Calculations

According to Scheme I above, MP can be converted to p-NP either directly or via the isomer, both pathways being catalyzed by the bifunctional-organoclay (DDMAEA). It is expected that due to the higher polarity of the phosphoryl group of the isomer, relative to the thio- phosphoryl group of MP, $k_2 \geq k_3$ (*18*).

The rate equations for scheme 1 are:

$$\frac{d[MP]}{dt} = -k_1[MP] - k_3[MP] \tag{1}$$

$$\frac{d[I]}{dt} = k_1[MP] - k_2[I] \tag{2}$$

$$\frac{d[p-NP]}{dt} = k_2[I] + k_3[MP] \tag{3}$$

The system of equations (1-3) was solved to get explicit expressions for the concentrations of the components (*15*). The values of the rate constants obtained by fitting the experimental data to the equations are presented in Table VIII.

Table VIII. Rate Constants (days-1) for the Hydrolysis of Methylparathion in the Presence of the DDMAEA-Clay at Different pH Values.

pH	k_1 MP $\longrightarrow$ I	k_2 I $\longrightarrow$ p-NP	k_3 MP $\longrightarrow$ p-NP
9	0.10	0.21	0.05
10	0.24	0.48	0.12
11	0.31	0.60	0.31
12	1.34	2.04	1.53

Catalytic Mechanism of the Bifunctional Organoclay

Whereas the catalytic effect of the long-chain bifunctional organoclay on the hydrolysis of organophosphate esters is very pronounced, the short-chain bifunctional organoclay (TMAEA) did not enhance the hydrolysis at all. The sorption of organic compounds to short-chain organoclays is, by and large, a Langmuir surface phenomenon, resulting in a conformational rigidity that may limit the degrees of freedom available to the sorbate at the surface, while the sorption of these compounds to long-chain organoclays resembles a partitioning process in that the organic substrate is incorporated into the 3-D organoclay matrix allowing considerable conformational flexibility. Small steric differences may have a profound effect on the catalytic process; it would seem that the incorporation of the substrate into the organoclay matrix is essential for the catalytic effect to take place. Incorporation of this nature can facilitate a proximity effect where the substrate and the catalytic group of the organoclay (the amino group) are brought close to each other by the adsorption process.

This is equivalent to increasing the concentrations of the reactants at the reaction site, which explains the enhanced reaction rates (*19, 20-22*). In the case of the short-chain bifunctional organoclay, due to the different type of sorption with its imposed higher rigidity, sorption may retard access at the proper orientation to the catalytic site rather than enhance it (as it does in the case of the long-chained bifunctional clay), and the proximity effect apparently does not exist.

According to the rate constants (Table VIII), at pH values of 9 and 10, approximately 65% of the reaction proceeds via the formation of the intermediate, and only 35% of the starting material hydrolyses directly to the final products ($k_1/k_3 \sim 2$). The intermediate isomer, in turn, hydrolyzes more rapidly than MP to the final products ($k_2 > k_3$). The mechanistic pathway of isomerization involves an O-CH$_3$ cleavage, which was reported in the past as a possible pathway for reactions involving organophosphate esters (*23*).

As the pH increases, the contribution of direct hydrolysis (k_3), which is also catalyzed by the bifunctional organoclay, increases and eventually exceeds that of the isomerization pathway (Table VIII). Therefore, one may conclude that the direct hydrolysis pathway is more sensitive to pH than the isomerization process. Overall, it is an interesting case of a catalytic reaction in which the pH not only affects the reaction rate, but also its path.

The experiments with the tertiary amino-LCOC were carried out in order to differentiate between 3 possible mechanistic pathways of the catalytic process as follows: a) a nucleophilic attack of the organoclay amino group on the reaction center of the organophosphate ester to form a transition state of a low activation energy, which in turn can be easily converted to the products; b) a similar attack of the organoclay amino group via a water molecule; and c) methylation of the amino group of the organoclay (due to a O-CH$_3$ cleavage of the MP molecule) to produce the isomer intermediate. A nucleophilic attack by a tertiary amine, is generally weaker than that by a primary amine, especially through pathway a, because of steric hindrances. Thus, the higher rate of the MP hydrolysis catalyzed by the tertiary amino-LCOC relative to that of catalysis by the primary amino-LCOC (the overall rate constant for the hydrolysis was 2.2 times faster in the presence of the bifunctional DDMAEDMA-organoclay), supports a general base catalysis pathway, or direct hydrolysis at higher pH values, and a methylation process to produce the isomer intermediate at the lower pH values.

Thiophosphoric acid esters, such as MP and tetrachlorvinphos, are hazardous pollutants, and their accumulation in the environment is a recognized ecological threat (*24*). Developing catalytic processes for their degradation is an urgent task of contemporary chemical technology and biotechnology. Due to the present relatively high cost of synthesizing the bifunctional organoclays, it is not likely that they will be used for application on a large scale. They can, however, be used to treat point source pollution, such as effluents of manufacturing lines in the pesticide industry.

The main importance of the present study is in demonstrating that organoclays that were formerly regarded mainly as excellent sorbents, can also be used as catalysts for various chemical reactions by incorporating functional headgroups capable of catalyzing a variety of reactions that are useful in chemical laboratories and industries and for environmental cleanup.

References

1. Qutubuddin, S.; Fu, X.A. In: *Nanosurface Chemistry,* Rosoff, M., Ed.; Marcel Dekker: NY, 2002; p 652.
2. LeBaron, P.C.; Wang, Z.; Pinnavaia, T.J. *Appl. Clay Sci.* **1999,** *15,* 11-29.
3. Giannelis, E.P. *Adv. Mater.* **1996,** *8,* 29-35.
4. Mortland, M.M.; Shaobai, S.; Boyd, S.A. *Clays Clay Miner.* **1986,** *34,* 581-585.
5. Boyd, S.A.; Lee, J.F.; Morland, M.M. *Nature* **1988,** *333,* 345-347.
6. Gibbons, J.J.; Soundararajan, R. *Am. Lab.* **1988,** *20,* 38.
7. Zielke, R.C.; Pinnavaia, T.J. *Clays Clay Miner.* **1988,** *36,* 403-408.
8. Gitipour, S.; Bowers, M.T.; Bodocsi, A. *J. Colloid Interface Sci.* **1997,** *196,* 191-198.
9. Lo, I.M.C.; Lee, S.C.-H.; Mak, R.K.-M. *Waste Manag. Res.* **1998,** *16,* 129-138.
10. Smith J.A.; P.R. Jaffe; Chiou, C.T. *Environ. Sci. Technol.* **1990,** *24,* 1167-1172.
11. Smith, J.A.; Galan, A. *Environ. Sci. Technol.* **1995,** *29,* 685-692.
12. Dentel, S.K. *Use of Organo-Clay Adsorbent Materials for Groundwater Treatment Applications;* Final Report submitted to Delaware State Water Research Institute, Newark, 1996.
13. Groisman, L.; Rav-Acha, Ch.; Gerstl, Z., Mingelgrin, U. *Appl. Clay Sci.* **2003,** *24,* 159-166.
14. Groisman, L.; Rav-Acha, C.; Gerstl, Z.; Mingelgrin, U. *J. Environ. Qual.* **2004,** *33,* 1930-1936
15. Rav-Acha, C.; Groisman, L.; Mingelgrin, U.; Tashma, Z.; Kirson, Z.; Sasson, Y.; Gerstl, Z. *Environ. Sci. Technol.* **2005,** (submitted for publication).
16. Mingelgrin, U.; Gerstl, Z. In *Organic Substances in Soil and Water: Natural Constituents and their Influence on Contaminant Behaviour,* Beck, A.J.; Jones, K.C.; Hayes, M.H.B.; Mingelgrin, U., Eds., Royal Society of Chemistry: Cambridge, 1993; pp 102-127.
17. Borisover, M.; Reddy, M.; Graber, E.R. *Environ. Sci. Technol.* **2001,** *35,* 2518-2524.
18. Faust, S. D.; Gomaa, Hm. *Environmental Letters* **1972,** *3,* 171-202.

19. Rav-Acha, Ch.; Ringel, I.; Sarel, S.; Katzhendler, J. *Tetrahydron* **1988**, *44*, 5879-5892.
20. Cang, H,; Brace, D. D.; Fayer, M. D. *J. Phys. Chem* **2001**, *B105*, 10007-10015.
21. Olmstead, E.G.; Harman, S.W.; Choo, P. L.; Cru, A. L. *Inorg. Chem.*, **2001**, *40*, 5420-5427.
22. Robertus, J. M.; Gebbink, K.; Martens, C. F.; Kenis, P. J. A.; Jansen, R. J.; Nolting, H.-F.; Sole, V. A.; Feiters, M.K.; Karlin, K.D.; Nolte, R. J. M. *Inorg. Chem.* **1999**, *38*, 5755-5768.
23. Chambers, J. A.; Levi, P. E. *Organophosphates. Chemistry, Fate, and Effects;* Academic Press, Inc: San-Diego, CA, 1992; pp 22-26
24. Kaloyanova, S.; Tarkowski, S. *Toxycology of Pesticides;* WHO: Kopenhagen, 1st Ed., 1981.

Biological Methods of
Subsurface Remediation

Chapter 6

Biological Treatment of Petroleum in Radiologically Contaminated Soil

C. J. Berry[1], S. Story[1], D. J. Altman[1], R. Upchurch[2], W. Whitman[2], D. Singleton[1], G. Plaza[3], and R. L. Brigmon[1]

[1]Savannah River National Laboratory, Westinghouse Savannah River Company, Aiken, SC 29808
[2]Departments of Microbiology and Ecology, University of Georgia, Athens, GA 30602
[3]Institute for Ecology of Industrial Areas, Katowice, Poland

This chapter describes *ex situ* bioremediation of the petroleum portion of radiologically co-contaminated soils using microorganisms isolated from a waste site and innovative bioreactor technology. Microorganisms first isolated and screened in the laboratory for bioremediation of petroleum were eventually used to treat soils in a bioreactor. The bioreactor treated soils contaminated with over 20,000 mg/kg total petroleum hydrocarbon and reduced the levels to less than 100 mg/kg in 22 months. After treatment, the soils were permanently disposed as low-level radiological waste. The petroleum and radiologically contaminated soil (PRCS) bioreactor operated using bioventing to control the supply of oxygen (air) to the soil being treated. The system treated 3.67 tons of PCRS amended with weathered compost, ammonium nitrate, fertilizer, and water. In addition, a consortium of microbes (patent pending) isolated at the Savannah River National Laboratory from a petroleum-contaminated site was added to the PRCS system. During operation, degradation of petroleum waste was accounted for through monitoring of carbon dioxide levels in the system effluent. The project demonstrated that co-contaminated soils could be successfully treated through bioventing and bioaugmentation to remove petroleum contamination to levels below 100 mg/kg while protecting workers and the environment from radiological contamination.

Introduction

The Savannah River Site (SRS), a Department of Energy (DOE) facility located in South Carolina, has generated non-hazardous petroleum and radiologically contaminated soils from spills and past disposal practices (1). The South Carolina Department of Health and Environmental Control (SCDHEC) regulations allow for burial of petroleum-contaminated soils in sanitary landfills with total petroleum hydrocarbon (TPH) concentrations below 100 mg/kg, but no allowances are made for disposal of radiologically and petroleum co-contaminated soil (2). Therefore, these co-contaminated soils were being stored in low-activity vaults for an indefinite period of time. SRS submitted a corrective action plan to SCDHEC that proposed *ex situ* cleanup of the petroleum portion of the soils using simple, inexpensive, and safe bioreactor technology (3). Final disposal of the treated soil, after treatment of the petroleum contamination, was buried in SRS trenches that accept low-level radiological wastes (4). The petroleum and radiologically contaminated soil (PRCS) bioreactor was developed to provide and demonstrate an efficient treatment pathway for this material to reduce operating costs, to provide a safe remedial method for treatment of spills, and to be applicable to other co-contaminated soils. In the present work, we describe how bacteria were isolated, screened and tested at the laboratory scale and applied to remediate tons of petroleum-contaminated soil in the PRCS bioreactor.

Biotreatment Technology

Biostimulation and/or bioaugmentation are effective alternatives to traditional physicochemical techniques for the cleanup of petroleum-contaminated soils (5). Current physicochemical techniques for disposal or decontamination of hydrocarbon-contaminated soils include landfill disposal, incineration, vapor extraction, detergent washing, and chemical oxidation (*6*). Biodegradation of petroleum hydrocarbons by stimulation of indigenous soil microorganisms, also known as biostimulation, is a proven remediation technology. Biostimulation involves the addition of electron acceptors, electron donors or nutrients to enhance the activity of indigenous microorganisms (*7*). Bioaugmentation involves the addition of indigenous or non-indigenous laboratory-grown microorganisms capable of biodegrading target contaminants (*7, 8*) or serving as donors of catabolic genes (*9*). Bioremediation of hydrocarbon-contaminated soils, which uses the natural ability of microorganisms to degrade and/or detoxify organic compounds, has been established as an efficient, economic, versatile, and environmentally sound treatment.

Impact of Radioactivity on Microorganisms

The impact of high levels of radiation on microbial activity and survival has been studied, but the impact of low-level waste on microbial survival and activity has not (*10*). Microbial survival studies have reported that *Bacillus* spores and *Kineococcus radiotolerans* have withstood radiation up to 3.5 kGy, and a ten percent survival of *Escherichia coli* was reported after a dose of 500 Gy (*11*). *Deinococcus radiodurans* (*12*) has survived a chronic dose of 20 kGy and an acute dose of 10 kGy. The lethal dose, or dose that would be expected to cause immediate incapacitation and death of a human within one week, is approximately 50 Gy (*13*). In general, bacteria are much more resistant to radiation fields than humans. Although strict dose rate levels are not used to define low-level radioactive waste, the petroleum-contaminated soils stored in the low-level vaults at the SRS do not generate doses greater than 100 µGy per year. Smith et al. (*14*) used risk-based modeling to assess the impact of disposal of radioactive petroleum waste in nonhazardous landfills and found that disposal of technologically enhanced, naturally occurring radiological materials presented a negligible risk to most potential receptors evaluated in their study. Since low-level waste storage is characterized based on risks to humans, any impact on microorganisms should be minimal.

Soil Treatment and Bioventing

Bioventing refers to enhanced bioremediation through the active or passive addition of oxygen (*15*). Enhanced bioremediation using bioventing to treat petroleum-contaminated soils requires an understanding of the basic principles of system design and microbial processes. When possible, contaminated soil is more efficiently treated if the biological treatment can be performed *ex situ* (*16*), since the addition of necessary nutrients, bulking agents, bacteria, and oxygen can be applied more easily than *in situ*. Biostimulation, or the addition of nutrients, can be applied both above ground in prepared beds or reactors and below ground using bioventing. Bioventing uses air injection or vacuum extraction to increase oxygen levels and is appropriate for relatively porous soil (*17*). However, contrary to soil vapor vacuum extraction, flow rates are relatively low to prevent stripping, but high enough to enhance microbial metabolism (*18*). Bioventing has been used to remediate gasoline-, diesel-, and PAH-contaminated soils (*19*).

Bioventing Requirements

If adequate amounts of oxygen, moisture and nutrients are available and the contaminants are accessible to the microorganisms, complete degradation of petroleum hydrocarbons can occur. Aerobic conditions and appropriate microorganisms are necessary for an optimal rate of bioremediation of soils contaminated with petroleum hydrocarbons (*17*). The oxygen content of soils depends on microbial activity, soil texture, water content, and depth. Low oxygen content/availability in soils has been shown to limit bioremediation of petroleum hydrocarbons (*20*). In a laboratory column experiment with acclimated soils, mineralization of hydrocarbons was severely limited when the oxygen content was below 10% (*21*).

Moisture levels are important for microbial enzymatic activity and proper operation of bioventing processes (*22*). In general, enzymatic reaction rates increase with increased moisture, although enzymatic reactions have been shown to decrease when specific metal ions were mobilized as a result of increased soil moisture (*23*). However, in a bioventing system, the presence of saturated soils limits airflow, permeability, or conductivity through the soil bed and impacts oxygen distribution. Soil moisture levels between 25% and 85% have been reported as suitable for bioremediation (*24*).

Microbial processes require nutrients for cellular processes, growth, and reproduction. Oxygen acts as an electron acceptor for aerobic bacteria and is required for cellular processes to occur. Nitrogen has been successfully introduced into the terrestrial subsurface for biostimulation using ammonia, nitrate, urea, and nitrous oxide (*25*). Several inorganic and organic forms of phosphate have been successfully used to biostimulate contaminated environments (*25*). In general, the addition of inorganic fertilizers in a ratio ranging from 9 to 600 to 1, carbon to nitrogen, potassium, and phosphorus has been reported to stimulate remediation of petroleum-contaminated soils (*6*).

Complex organic sources of nutrients, e.g., compost, have also been shown to increase microbial activity (26) and diversity that can enhance bioremediation of organic contaminants (27).

Bioavailability of Contaminants

The bioavailability of contaminants is an important factor in bioremediation. The bioavailability of a chemical may be described by its mass transfer rate relative to its uptake and degradation rates by microorganisms (28). If the capacity for hydrocarbon degradation is present and environmental conditions are amenable, the microorganisms must have access to the contaminants for degradation (29). Reduced bioavailability could be caused by low aqueous solubility and strong sorption to soils or sediments (30). It has been shown that the water-dissolved fraction of chemicals is more available to soil microorganisms (31). The use of surfactants has been shown to increase biodegradation of hydrocarbon contaminants by increasing bioavailability (32).

Temperature is an important parameter for most bioremediation sites because of its impact on the availability of contaminants and the activity of the microorganisms. Especially true in northern latitudes, seasonal variation can also impact bioremediation sites (33). For optimal contaminant removal, biological treatment of organic pollutants such as petroleum-based hydrocarbons is performed at moderate temperatures (20° to 37°C) in order to increase metabolic activity, diffusion, and mass transfer.

Modifying soil composition and structure through mechanical means or amendments can significantly influence bioremediation activities. Bulking agents are materials of low density that lower soil bulk density, increase porosity, moisture retention and oxygen diffusion, and can help to form water-stable aggregates increasing aeration and microbial activity (34). Indigenous microbes, those growing naturally in soil, sediment, or groundwater, and non-indigenous microbes, those added from an external source, have been used in bioremediation of petroleum hydrocarbons (35). However, refining of petrochemicals results in the generation of oil sludge consisting of hydrophobic compounds resistant to biodegradation (36). The addition of surfactant-producing non-indigenous microbes or synthetic surfactants has been used in soil treatment to help increase availability of these recalcitrant materials (37). Moreover, the production and presence of biosurfactants has been shown to have many of the benefits of synthetic surfactants, as well as being biodegradable and nontoxic (38). Although non-indigenous organisms must be able to compete for nutrients and retain their ability to degrade contaminants, bioaugmentation has been shown to work in field conditions for a variety of organic compounds (39).

Methods

Microbial Isolation and Characterization

A consortium of microbes isolated at the Savannah River National Laboratory (SRNL) from a petroleum-contaminated site was added to the PRCS system. The organisms were isolated from sludge samples obtained from a 100-year-old oil refinery near Czechowice-Dziedzice, Poland (40). The aged sludge was acidic (pH 2) and composed of asphaltics that were highly contaminated with polycyclic aromatic hydrocarbons (PAHs) (41).

One gram samples (wet weight) of sludge or biopile material were suspended in 10 ml of 0.1% (w/v) sodium pyrophosphate buffer (pH 7) and vortexed. Serial dilutions were on plates with minimal agar (pH 4) exposed to naphthalene vapor for two weeks. Bacterial colonies with distinct morphotypes were picked and transferred to the same agar medium for purification and subsequent characterizations. The number of potential phenanthrene degraders was determined by spraying a saturated solution of phenanthrene in hexane directly to the colonies on the agar surface. After an additional week of incubation, colonies that removed the phenanthrene crystals around their periphery were selected and characterized further. Microorganism were then identified as previously described (41).

The biosurfactant exudate was evaluated for each isolate and those determined to have a surface tension-altering property consistent with a surfactant were retested. In preparation for addition to the bioreactor, microbial isolates were grown in peptone, tryptone, yeast, and glucose (PTYG) medium. The PTYG media consisted of 1 g/L of peptone, 1 g/L of tryptone, 2 g/L of yeast, 1 g/L of glucose, 0.45 g/L of $MgSO_4$, and 0.07 g of $CaCl_2$ (all reagents from Fisher Scientific or Difco-Becton, Dickenson and Company, Franklin Lakes, NJ). Isolates were grown at 28°C on a shaker flask until bacterial densities were greater than 1 x 10^7 cell/ml. Active cultures were in log phase growth when 2 liters were prepared for direct addition to the PCRS bioreactor.

Reactor Construction

The bioreactor was constructed from a 6.75 yd^3 volume skid-pan by adding a false floor, sample ports, lid with seals, gauge ports, HEPA filters, and air pumps (10).

Bioreactor Soil Sampling

Soil sampling required protection from radioactive contamination. This included wearing multiple sets of gloves, using hand-held radiological monitoring equipment, and swipes for alpha contamination (Eberline AC-3, Thermo Electron Corporation, San Jose, CA), beta/gamma contamination (Ludlum model 12 with an HP 110 probe), and radiation (RO-20, Thermo Electron Corporation). Before sampling, all system pumps were turned off and all relief valves were opened. Once the pressure/vacuum gauges on the PRCS bioreactor read "zero" the access port was opened. Soil samples were taken by hand using a three-foot carbon steel sampling rod with a stainless steel sampling probe. Multiple 50 gram soil samples were taken from randomly selected holes to screen the entire vertical soil profile of the PRCS system. The soil from each hole was immediately placed in a sterile, 50 ml polypropylene centrifuge tube (Corning, Acton, MA). Once sampling was complete, the samples were packed on ice and transported in to an SRS radiological facility located in the SRNL. Collected soils were stored at room temperature, and analyses were performed within 7 days of sampling.

Measurements

Hydrocarbon concentrations were performed using a gravimetric method, and analyses were performed using gas chromatography in conjunction with a mass selective detector (10). Soil nutrient levels, pH, and soil moisture were monitored by quantitating the water-soluble inorganic forms of nitrate, nitrite, ammonia, potassium, and phosphate (10). Gas samples were taken from a sampling valve downstream of the effluent line HEPA filter were analyzed in a non-radiological laboratory (10) for carbon dioxide, oxygen and hydrocarbons. Carbon dioxide generation rates were determined on weathered compost as previously described (10) using less weathered but similar material from the same source. Temperature, vacuum, and air flow rates were taken from the unit (10).

Reactor Loading

Two radiological material storage boxes (B-12 boxes) containing 7,340 lb of petroleum and radioactive contaminated soil were loaded onto the grating inside the bioreactor. A Typar® -style (Tri-State Stone® & Building Supply, Inc., Bethesda, MD) geotextile fabric was placed on top of the grating inside the bioreactor. While loading, the soil was amended by mixing in 6 ft^3 of compost,

1.36 pounds of ammonium nitrate (Fisher Scientific) and 0.54 pounds of 10-10-10 fertilizer (Lowe's®, Aiken, SC). Bags of mixed compost and fertilizer were manually added to the reactor system toward the end of each B-12 transfer. Once the bioreactor was filled, the soil inside the reactor was leveled using a hoe. When level, the soil covered the lower two rows of temperature and pressure gauges. The upper third row of gauges, vacuum relief valve, and pressure relief valve were not in contact with the soil. After leveling, an estimated total of 80 gallons of water was added to the system. The initial contamination level in the system, fully loaded, was estimated to be 25,000 mg/kg TPH, based on analyses of the loaded soil.

Operation of the Reactor

The PRCS bioreactor operated for 22 months in various configurations treating the contaminated soil. Initial soil TPH concentration was greater than 25,000 mg/kg, and the final TPH concentration was 45 mg/kg. Ten days after loading and staging the PRCS system, the system began continuous operation. System parameters were initially adjusted so that there was a slight vacuum, less than 0.6 inch water, on all of the pressure gauges. The system operated in a variety of configurations and two different locations during testing. The following operating ranges were used as guidance for operation of the bioreactor: oxygen concentration, 10-21% in air; soil moisture, 8-20% by weight; carbon:nitrogen:phosphorus ratio, 100:10:2 or greater for nitrogen and phosphorus; soil pH, 4-7; and soil temperature, 15-35 °C.

The system operated continuously, except for down times due to sampling, during the first four months of operation. During month five, 58 gallons of tap water, open to the atmosphere for 5 days, were pumped into the system to adjust soil moisture levels, and three soil samples were pulled from the reactor. During the sixth month of operation, temperature and carbon dioxide production levels were low, and the inlet and outlet air pumps were turned off. The system was checked biweekly to operate the pumps and check carbon dioxide levels. In months eight to eleven, the pumps were operated periodically to purge carbon dioxide from the system and provide oxygen. During the tenth month of operation, the system was moved from a covered facility with power to an outside radiological storage area. A cover was constructed to provide protection from the sun and rain, and the system was operated using a portable generator.

During month twelve, the inlet pumps were replaced with larger flow pumps to facilitate increased aeration and carbon dioxide removal. Carbon dioxide removal was achieved by operating the system for approximately 90 mm on a weekly basis. In the thirteenth month of operation, a solar-powered pump system was installed. During month nineteen a soil sample was taken for analyses by General Engineering Laboratories (Charleston, SC). Analyses of PRCS soil showed TPH concentrations of 279 mg/kg. This was above the 100 mg/kg required by SCDHEC for disposal. Carbon dioxide was not measured exiting the system in month twenty-two. The system soil was sampled in month twenty-two for external analyses. Results were obtained in month twenty-two and confirmed successful treatment of the soil to less than 100 mg/kg. The soil was then disposed in a lined slit trench located on the low-level radiological burial ground, E-Area, at SRS.

Soil Analyses

Soil samples were taken from one of the B-12 boxes prior to loading the system and from the PRCS five times during operation. Samples were taken at three weeks, four months, fourteen months, nineteen months, and twenty-one months. Soil hydrocarbon concentrations were measured for all of the sampling events. Soil moisture levels were measured at months four and fourteen, and soil pH was measured on the B-12 box during week three. Soil nutrient levels were obtained from the B-12 box prior to loading the system, at three weeks and fourteen months.

Results and Discussion

Twelve of the bacterial isolates were shown to have consistent activity for bioremediation of petroleum compounds. Nine of these organisms added for bioaugmentation are listed in Table I. Isolates 1-3, *Alcaligenes piechaudii* SRS, *Ralstonia pickettii* SRS, and *Pseudomonas-putida* Biotype B SRS, all demonstrate the ability to produce biosurfactants in the presence of petroleum compounds, the formation of which was noted during culturing conditions (Table I). Isolates 4-9 all demonstrate the ability to biodegrade a variety of petroleum hydrocarbons (Table I).

System Monitoring

Soil moisture analysis was performed on samples pulled during months four and fourteen. In month four, the average soil moisture level from 5 samples was 10.6%, which was close to the lower operating limit of 8%. Additional water was added to the system. In month fourteen the average soil moisture levels from 3 samples was 12.4%. Soil moisture levels were indirectly monitored by measuring the level of water in the bottom of the PRCS system using a stud

Table I. Bacteria Cultures Used for Bioaugmentation

Number	Isolate	Identification
1	CZOR-LIB (KN-1)	*Alcaligenes-piechaudii* SRS
2	BP-20 (KN-2)	*Ralstonia pickettii* SRS.
3	CZOR-LlBsm (KN-3)	*Pseudomonas putida* Biotype B SRS
4	BPB	*Flexibacter cf. sancti* SRS
5	BPC	*Pseudomonas fredriksbergensis* SRS
6	BPE	*Staphylococcus warneri. LMG 19417* SRS
7	BPF	*Sphingomonas* SRS
8	BPH	*Sphingomonas Sp. S37* SRS
9	BPI	*Phylobacterium* SRS
10	CZOR-L1B (KN-1)	*Alcaligenes piechaudii* SRS - (α Proteobacterium TA-Al)

finder. Greater than one inch of water was detected in the bottom of the reactor until month nineteen.

Soil pH levels were measured in soils pulled from the B-12 box prior to system startup and at three weeks into operation. Soil pH was 5.9 in the B-l2 box sample initially and 6.2 during week three. Soil nutrients that were measured using ion chromatography included nitrate, nitrite, ammonia, phosphate, and potassium. Analyses were performed on the B-l2 box prior to loading and on the three-week soil samples. Although total nitrogen was not determined due to the high levels of ammonium present, available nitrogen, potassium, and phosphate concentrations were present.

Due to the heterogeneity of the samples taken in the third week, all sampled material was combined, mixed by hand, and separated into four samples. The samples analyzed by the outside laboratories were analyzed once. Results from month four and fourteen are averages of duplicate samples. Results for the SRNL analyses of the B-12 box, four month and fourteen month samples are from gravimetric analyses.

General Engineering Laboratories also analyzed soil samples for benzene toluene, ethyl benzene, and xylenes (BTEX) using SW-846 8260B; PAHs using SW-846 8270; and GRO using SW-846 8015B (*42*). No detectable BTEX, PAH or gasoline range organics were measured. Final analyses by Accura Analytical, (Norcross, GA) demonstrated TPH, as measured by diesel range organic analysis, was 45 mg/kg, which was less than the 100 mg/kg maximum disposal level required by SCDHEC at month twenty-two.

Flow rates entering and exiting the system were primarily determined by the pump type being used. Three sets of pumps were used to operate the system. Flow rates using the medium-sized pumps ranged from 30 to over 75 SCFH with an average flow rate of 48 SCFH. Inlet and outlet flow rates using the larger pumps averaged 225 SCFH. Inlet and outlet flow rates using the solar-powered pumps ranged from 10 to 20 SCFH with an average flow rate of 17.5 SCFH.

PRCS bioreactor temperatures were taken at three points, one in the headspace of the system and two in the soil profile of the unit. Maximum soil temperatures, above 25°C, were measured during the first four months of operation and during months eleven through fifteen. Minimum temperatures, below 15°C, were measured at the end of the fifth month of operation through the ninth month and in months seventeen through twenty-one. Soil temperatures changed with the median outdoor temperature as the cell was sheltered from rain. Methane was not detected in any sampling of the system using the portable analyzer or in any gas bag samples. VOCs also were not detected in any gas bag samples. Oxygen concentrations were measured during the first three months of operation averaging value was 17.12%.

Carbon dioxide measurements were used to monitor hydrocarbon degradation, microbial activity, and to indicate when the system had completed bioremediation of the contaminated soil (Figure 1). Carbon dioxide concentrations exiting the system were used with system flow parameters, the mass of contaminated soil, and the stoichiometric ratio of an alkane, $> C11$, to carbon dioxide in a hydrocarbon oxidation reaction to calculate theoretical hydrocarbon degradation rates.

The major physical impacts on reactor performance were soil temperature and pump operation. Soil temperatures changed with ambient temperature changes (probability $> t < 0.0001$). TPH degradation and soil temperature also were also somewhat related (probability $> t < 0.0001$) (Figure 1). During the first decrease in temperature, days 123-250, carbon dioxide production dropped below 10 mg/kg/day. Carbon dioxide production also indicated a general decrease in degradation rate during the second temperature decrease, days 500-600, although overall degradation rates were higher during the second temperature decrease.

Carbon dioxide production was also reduced during day 175 through day 400 when pump operation was intermittent (Figure 1). Low TPH degradation was observed while soil temperatures were relatively high, above 15°C, but flow rate through the system was low. When continuous flow was returned to the system using the solar pumps, TPH degradation increased to over 50 mg/kg/day, but then decreased throughout the remainder of system operation. Overall, pump

98

operation was related to TPH degradation (probability > t = 0.4), as shown in Figure 1.

Using a portable generator reduced the amount of oxygen (air) that was added to the PRCS system and impacted TPH degradation rates compared to continuous operation with permanent power or with solar pumps (Figure 1).

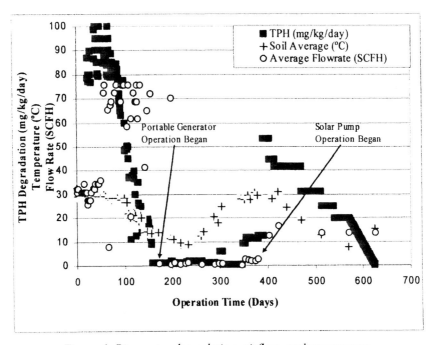

Figure 1. Bioreactor degradation, airflow, and temperature.

During operation in the field using a portable generator, the bioreactor operated an average of once a week. The system was designed to vent to the atmosphere through the HEPA filters. Therefore, the volume expansion resulting from temperature increases or net gas/vapor production would cause gaseous material to be vented from the system and would change the headspace volume used to calculate TPH degradation.

In general, using carbon dioxide concentrations to monitor TPH degradation has uncertainties, but is an effective parameter to use for external measurements. Using the mass ratio between an alkane and carbon dioxide in the hydrocarbon oxidation reaction yields the following relationship:

$$-r_{TPH} \approx 0.31 \frac{mgTPH}{mgCO_2} r_{CO2} \qquad (1)$$

where $-r_{TPH}$ is the rate of hydrocarbon degradation and r_{CO2} is the carbon dioxide production rate. TPH degradation is underestimated when aromatics, PAHs, and olefins are being degraded, but it is relatively accurate for branched alkanes. The typical hydrocarbon content of gasolines is 25-40% isoalkanes, and 20-50% aromatics (43). Huesemann (44) showed the general rate of biodegradation of petroleum compounds was in the order of n-alkanes > branched-chain alkanes > branched alkenes > low-molecular-weight n-alkyl aromatics > monoaromatics> cyclic alkanes > polynuclear aromatics > asphaltenes. Equation (1) would accurately estimate TPH degradation during early biodegradation but would overestimate degradation rates as treatment continued. Factors ignored in Equation (1) include the impact of the soil organic content, amendments added to the system, and the volatilization of breakdown products besides carbon dioxide. SRS soils are generally sandy with low organic carbon content so the impact of background organic carbon contributing to carbon dioxide generation should be minimal. Composted materials were also added to the bioreactor and contributed to carbon dioxide production but were neglected based on low production rates (10). Using carbon dioxide to monitor system operation was useful especially since internal system sampling was limited due to radiological protection issues. Although TPH degradation rate estimates are impacted by many factors, monitoring carbon dioxide production provided a straightforward monitoring tool during operation of the PRCS bioreactor.

Examination of GC/MS sample runs on untreated and treated soil showed that most of the petroleum contamination consisted of an unresolved complex mixture of hydrocarbons. This mixture has been described as resulting from the chromatographic overlap of thousands of compounds (45). Analyses of the total ion chromatograph did not reveal the presence of distinct chemical compounds but did show a mixture of co-eluting compounds with column residence times falling after the internal standard, deuterium-labeled anthracene. The total ion chromatogram was examined for general trends and specific PAH masses. Generally, mass per charge responses increased by 14 units in the unrefined area, indicating an additional carbon group (46). Major masses consistent with PAHs were not identified, and NIST library searches, with a probability greater than 50, did not identify any specific compounds. The extended storage period for this material before processing probably contributed to the small number of compounds that were identified using GC/MS analyses. Easily degradable and identifiable compounds were preferentially degraded first (44), probably during storage.

Degradation of this complex mixture of compounds required microorganisms with specialized enzymatic activities. The phenomenon has been described (35), and degradation of this complex mixture of compounds is slower than degradation of alkanes. The organisms that were inoculated into the PRCS system were isolated from a refinery site containing asphaltenes and other complex materials. Enzyme systems that were present in microorganisms in this refinery site waste could use these more complex compounds for growth. Although specific organisms and specific enzyme activity were not tracked in this investigation due to the radioactive nature of the waste, a change in the active microbes during PRCS operation is probable. The addition of select isolates that produce biosurfactants is believed to have enhanced remediation through several different mechanisms. The production of the biosurfactant increases the biological availability of PAHs and other hydrophobic petroleum compounds. As such, the natural ability to produce surfactants increases the efficiency of the microorganisms to degrade and metabolize PAHs in the weathered, petroleum-contaminated soils.

Using bioventing to treat petroleum-contaminated soil is a well-documented approach (47), but using bioventing to treat radiologically and petroleum co-contaminated soils has not been reported. During this testing it was demonstrated with proper engineering controls for worker and environmental protection, co-contaminated soils can be safely biovented. Bioventing in this bioreactor was designed to maximize the biodegradation of petroleum contaminants with little volatilization. Most of the common soil environmental radioactive contaminants (i.e., cesium, plutonium, and uranium) found at SRS have low volatility so release of radioactive material out of the system would not be expected. To ensure this did not occur, HEPA filters were placed on all process entry and exit points to trap any particulates, and the system was operated under a slight vacuum to protect the environment. Finally, the system was shut down, monitored, and protective clothing was worn when the bioreactor was opened. Based on the biodegradation rates and complete treatment in the PRCS bioreactor, the impact of the radiological contamination on the treatment was minimal.

Conclusions

Based on the results of this study the following conclusions are offered. A biovented bioreactor system can be used to effectively treat future co-contaminated soils at SRS and other locations. Carbon dioxide measurements were shown to be a good indicator and monitoring tool for microbial activity and TPH degradation. Soil temperature and the oxygen supply were identified as two important parameters that control the rate of biodegradation. Complete treatment time required twenty-two months of operation and reduced TPH levels from over 20,000 mg/kg to 45 mg/kg. Soil was permanently disposed of in low-level Rad waste trenches resulting in a significant reduction in disposal costs. Both carbon dioxide and hydrocarbon should be monitored *in situ* to determine the extent and rate of TPH degradation. The use of heating strips or the addition of insulation should be considered when treating soil in temperate climates. Biological treatment of co-contaminated soil was successfully completed using an *ex situ* bioreactor system safely.

References

1. Lombard, K.; Hazen, T.C. *Test Plan for the Soils Facility Demonstration - A Petroleum Contaminated Soil Bioremediation Facility;* WSRC-RP-94-0179; DOE: Savannah River Site; 1994.
2. SCDHEC. *South Carolina Hazardous Waste Management Act;* #61-107.18; 2001.
3. Kastner, J.R.; D. J. Altman; C. B Walker. *Corrective Action Plan - Degradation of Petroleum Contaminated Soils Using Ex-Situ Bioreactors and Bioventing,* WSRC-RP-97-25; Savannah River Site; 1998.
4. Mamatey, A.E., (ed.). *Savannah River Site Environmental Report for 2003,* WSRC-TR-2004-00015; DOE: Savannah River Site; 2003.
5. Dua, M.; Singh, A.; Sethunathan, N.; Johri, A.K. *Appl. Microbiol. Biotechnol.* **2002,** *59,* 143-152.
6. Riser-Roberts, E. *Remediation of Petroleum Contaminated Soils: Biological, Physical, and Chemical Processes;* Lewis Publishers CRC Press LLC: Boca Raton, FL, 1998; p 542.
7. Widada, J.; Nojiri, H.; Omori, T. *Appl. Microbiol. Biotechnol.* **2002,** *60,* 45-59.
8. Vogel, T.M. *Curr. Opin. Biotechnol.* **1996,** *7,* 311-316.
9. Top, E.M.; Spningael, D.; Boon, N. *FEMS Microbiol. Ecol.* **2002,** *42,* 199-208.
10. Berry, C.J. Masters thesis; Georgia Institute of Technology, Atlanta, GA, **2005.**
11. Phillips, R.W.; Wiegel, J.; Berry, C.J.; Fliermans, C.; Peacock, A.D.; White, D.C.; Shimkets, L.J. *International Journal of Systematic and Evolutionary Microbiology.* **2002,** *52,* 933-938.
12. Anderson, A.W.; Nordan, H.C.; Cain, R.F.; Parrish, G.; Duggan, D. *Food Technol.* **1956,** *10,* 575-577.
13. Charpak, G., R. L. Garwin. *Europhysics News.* **2002,** *33,* 24.

14. Smith, K.P.; Arnish, J.J.; Williams, G.P.; Blunt, D.L. *Environ. Sci. Technol.* **2003**, *37*, 2060-2066.
15. DuPont, R.R. *Environ. Prog.* **1993**, *12*, 45-53.
16. Alexander, M. *Biodegradation and Bioremediation;* Academic Press: San Diego, CA, 1999; Vol. l, p453.
17. USEPA. *Manual: Bioventing Principles and Practice; 540/R 951534a* U.S. EPA, Washington, DC; 1995.
18. Hinchee, R.E.; Downey, D.C.; Dupont, R.R.; Aggarwal, P.K.; Miller, R.N. *J. Hazard Mater.* **1991**, *27*, 315-325.
19. Miller, R.V.; Poindexter, S. *Strategies and Mechanisms for Field Research in Environmental Bioremediation;* American Academy of Microbiology, Washington, D.C. 1994; p 20.
20. von Wedel, R.T.; Mosquera, S.F.; Goldsmith, C.D.; Hater, G.R.; Wong, A.; Fox, T.A.; Hunt, W.T.; Paules, M.S.; Quiros, J. M.; Wiegand, J.W. *Water Sci. Technol.* **1988**, *20*, 501-503.
21. Freijer, J.I. *J. Environ. Qual.* **1986**, *25*, 296-304.
22. Dick, W.A.; Tabatabai, M.A. *Use of Immobilized Enzymes for Bioremediation. in Bioremediation of Contaminated Soils;* American Society of Agronomy Inc., Madison, WI, 1999; pp 315-338
23. Acosta-Martinez, V. and Tabatabai, M.A. *Soil Biol. Biochem.* **2001**, *33*, 17-23.
24. Alexander, M. *Introduction to Soil Microbiology;* John Wiley & Sons: New York, 1977; Vol. 1, p 467.
25. USEPA, *Bioremediation of Hazardous Waste Sites Workshop, CERI 89-1-1;* U.S. EPA: Washington, DC, 1989.
26. Shimp, R.J.; Pfaender, F.K. *Appl. Environ. Microbiol.* **1985**, *49*,402-407.
27. Zhou, J.; Xia, B.; Treves, D.S.; Wu, L.-Y.; Marsh, T.L.; O'Neill, R.V.; Palumbo, A.V., Tiedje, J.M. *Appl. Environ. Microbiol.* **2002**, *68*, 326-334.
28. Borsma, T.N.P.; Middeldorp, P. J. M.; Schraa, G.; Zehnder, A. J. B. *Environ. Sci. Technol.* **1997**, *31*, 248-252.
29. Villemur, R.; Déziel, R.E.; Benachenhou, A.; Marcoux, J.; Gauthier, E.; Lépine, F.; Beaudet, R., and Comeau, Y. *Biotechnol. Prog.* **2000**, *16*, 966-972.
30. Harms, H. and Bosma, T.N.P. *J. Ind Microbiol. Biotechnol.* **1997**, *18*, 97-105.
31. Thomas, J.M.; Yordy, J.R.; Amador, J.A.; Alexander, M. *Appl. Environ. Microbiol.* **1986**, *52*, 290-296.
32. Bruheim, P.; Bredholt, H.; Eimhjellen, K. *Can. J. Microbiol.* **1997**, *43*, 17-22.
33. Ward, D.M.; Brock, T.D. *Appl. Environ. Microbiol.* **1975**, *31*, 764-772.
34. Hillel, D. *Soil Structure and Aggregation: Introduction to Soil Physics;* Academic Press: London, 1980; pp 40- *52*, 200-204.

35. Atlas, R.M. *Microbiological Review.* **1981**, *45,* 180-209.
36. El-Nawawy, A.S.; El-Bagouri, I.H.; Abdal, M.; Khalafai, M.S. *World Journal of Microbiology and Biotechnology.* **1992**, *8,* 618-620.
37. Roane, T.M.; Josephson, K.L.; Pepper, I.L. *Appl. Environ. Microbiol.* **2001**, *67,* 3208-3215.
38. Makkar, R.S., and K. J. Rockne. *Environmental Toxicology and Chemistry.* **2003**, *22,* 2280-2292.
39. Barbeau, C.; Deschenes, L.; Karamanev, D.; Comeau, Y.; Samson, R. *Appl. Environ. Microbiol.* **1997**, *48,* 745-752.
40. Altman, D.J.; Hazen, T.C.; Tien, A.J.; Worsztynowicz, A.; Krzysztof, U. *The Czechowice Oil Refinery Bioremediation Demonstration of a Process Waste Lagoon – Czechowice-Dziedzice. Poland;* WSRC RP-97-214, Westinghouse Savannah River Company: Aiken, S.C.,1997.
41. Brigmon, R.L.; Berry, C.J.; Story, S.; Altman, D.; Upchurch, R.; Whitman, W.B.; Singleton, D.; Plaza, G.; Ulfig, K. *Bioremediation of Petroleum and Radiological Contaminated Soils at the Savannah River Site: Laboratory to Field Scale Applications;* WSRC-MS-2004-00363, DOE: Savannah River Site, 2004.
42. USEPA. *SW-846 On Line: Test Methods for Evaluating Solid Wastes Physical Chemical Methods.* **2005** [cited 2005]; Available from: http://www.epa.gov/epaoswer/hazwaste/test/main.htm.
43. IARC. *Diesel and Gasoline Engine Exhausts and Some Nitroarenes;* International Agency for Research on Cancer: Lyon, France, 1989; Vol. 46, pp 458.
44. Huesemann, M.H. *Environ. Sci. Technol.* **1995**, *29,* 7-18.
45. Frysinger, G.S.; Gaines, R.B.; Xu, L.; Reddy, C.M. *Environ. Sci. Technol.* **2003**, *37,* 1653-1662.
46. Prince, R.C.; Grossman, M.J. *Appl. Environ. Microbiol.* **2003**, *69,* 5833-5838.
47. Leson, G.; Winer, A.M. *Journal of the Air and Waste Management Association.* **1991**, *41,* 1045-1054.

Chapter 7

Natural Attenuation of Acid Mine Drainage by Acidophilic and Acidotolerant Fe(III)- and Sulfate-Reducing Bacteria

Sarina J. Ergas[1], Jaime Harrison[1], Jessica Bloom[2],
Kristin Forloney[3], David P. Ahlfeld[1], Klaus Nüsslein[3],
and Richard F. Yuretich[2]

Departments of [1]Civil and Environmental Engineering, [2]Geosciences,
and [3]Microbiology, University of Massachusetts, Amherst, MA 01003

Natural attenuation of acid mine drainage (AMD) was
investigated at Davis mine, a long-abandoned pyrite mine in
Western Massachusetts. Studies at both flask and field scales
included monitoring of groundwater hydrology and chemistry,
microbial community analysis and flask microcosm studies.
The results showed that mixing of contaminated groundwater
with water from higher topographies and dissolution of silicate
minerals restricts the contamination to a narrow region of the
site. Microbial Fe(III) reduction appears to be widely
distributed throughout the site, while sulfate reduction is
observed in higher pH zones, peripheral to areas of highest
contamination.

Introduction

Acid mine drainage (AMD) is caused when sulfidic ores (primarily pyrite, FeS_2) are exposed to oxygen and moisture through mining activities. AMD sites are characterized by high levels of acidity, elevated concentrations of Fe(III), and an excess of heavy metals. This combination of pollutants is toxic to almost all forms of aquatic life; only a few hardy plants and specialist microorganisms are able to survive in AMD streams (*1*). A number of treatment processes have been used to remediate AMD sites including, anoxic limestone drains, constructed wetlands and bioreactor systems (*2*). Natural attenuation, however, can be a low-cost alternative to engineered treatment processes for AMD sites if health and environmental risks are low. The central goal of this research is to understand the role that geochemical, hydrological and biological processes play in the natural attenuation of AMD. Research is being conducted at both the flask and field scales at an AMD site including:

- Field measurements of shallow and bedrock groundwater geochemistry and hydrogeology to identify zones of sulfate and Fe(III) reduction,
- Investigation of microbial diversity in field and microcosm samples using molecular techniques based on nucleic acid assays of the 16S ribosomal RNA,
- Microcosm studies to examine the effect of organic substrate availability and pH on microbial Fe(III)- and sulfate-reducing activity in different regions of the AMD site.

AMD Generation

Generation of AMD involves a cycle of biological and abiotic reactions that have been described in detail (*1, 2, 3*). The first step occurs when pyrite ores are exposed to oxygen and moisture:

$$2FeS_2 + 7O_2 + 2H_2O \rightarrow 2Fe^{2+} + 4SO_4^{2-} + 4H^+ \qquad (1)$$

This reaction can occur abiotically; however, the presence of chemolithotrophic bacteria have been shown to speed the reaction rate by up to six orders of magnitude (*2*). Fe(II) produced in Reaction 1 can be used as an electron donor for acidophilic chemolithotrophic bacteria, such as *Acidithiobacillus ferrooxidans*, at an optimal pH of around 2:

$$4Fe^{2+} + O_2 + 4H^+ \rightarrow 4Fe^{3+} + 2H_2O \qquad (2)$$

The Fe(III) produced in Reaction 2 can be hydrolyzed, yielding further acidity and ferric hydroxide precipitate:

$$4Fe^{3+} + 12H_2O \rightarrow 4Fe(OH)_3 + 12H^+ \qquad (3)$$

An alternative fate for Fe(III) is further pyrite oxidation, which occurs under either aerobic or anaerobic conditions:

$$FeS_2 + 14Fe^{3+} + 8H_2O \rightarrow 15Fe^{2+} + 2SO_4^{2-} + 16H^+ \qquad (4)$$

Fe(II) and acidity produced in Reaction 4 fuels further ferrous iron oxidation (Reaction 2) setting up a self-perpetuating cycle of AMD generation until some factor, such as pyrite, is depleted.

Natural Attenuation of AMD

Abiotic processes attenuating acidity and heavy-metal concentrations in water leaving AMD sites are dilution of mine drainage with uncontaminated water, precipitation-sedimentation, adsorption-coprecipitation and neutralization (4). Mixing of ambient groundwater and surface water from unaffected tributaries with AMD dilutes the concentrations of dissolved species and raises the pH. The solubility of Fe(III) decreases at high pH and Fe(II) oxidizes to Fe(III), diminishing the total iron content of the water. A similar rise in pH results from the reaction of AMD with surrounding bedrock. Although carbonate lithologies are most effective at neutralizing acidity, weathering of silicate minerals in mining waste rock or bedrock can also contribute alkalinity over a long time period (5).

The main biological processes that reverse acidification caused by AMD are denitrification, methanogesis, and Fe(III) and sulfate reduction. Given the high concentrations of sulfate and ferric iron present at AMD sites, Fe(III) and sulfate reduction are expected to dominate. Sulfate-reducing bacteria constitute a diverse group of heterotrophic anaerobes with the common ability to conserve energy via dissimilatory reduction of sulfate to hydrogen sulfide (6):

$$CH_3COO^- + SO_4^{2-} + H^+ \rightarrow H_2S + 2HCO_3^- \qquad (5)$$

This process increases the surrounding pH, and the resulting products allow the precipitation of metal sulfides and aluminum hydroxide.

Over the last decade a number of studies have demonstrated that microbial Fe(III) reduction also plays an important role in AMD neutralization, by generating alkalinity and decreasing the concentration of Fe(III) available for further pyrite oxidation (7, 8):

$$CH_2O + 2Fe_2O_3 + 3H_2O \rightarrow 4Fe^{2+} + HCO_3^- + 7OH^- \qquad (6)$$

Davis Mine

Our study investigates the Davis Mine in Rowe, Massachusetts (Figure 1). Once the largest and most productive pyrite mine in Massachusetts, the Davis Mine has been abandoned for 90 years, and no effort to remediate the site has ever been undertaken. Approximately 3 ha of exposed waste rock, containing significant amounts of pyrite, lies south of the mine shaft and is largely devoid

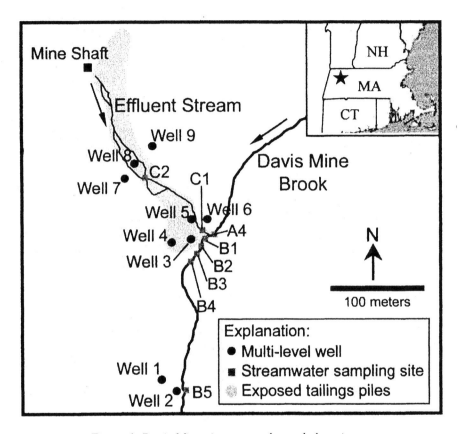

Figure 1. Davis Mine site map and sample locations.

of plant life, although the area around the site is now re-forested with a mixed hardwood community. The regional geology consists of metamorphic schist and gneiss, part of the Devonian-age Hawley Formation. In the vicinity of Davis Mine, waste rock and bedrock mineralogy consists primarily of silicate minerals, such as quartz, albite, chlorite and hornblende, and no carbonate minerals. The area is overlain by glacial till typical of upland New England. AMD contaminates nearby Davis Mine Brook, which no longer contains fish or other macrobiota.

Data from ground and surface water chemistry measurements at the mine site over more than 20 years indicates Davis Mine is in a state of dynamic equilibrium between acid generation and natural attenuation processes (9). The long period of stability and small areal extent provided an opportunity to evaluate the biogeochemical processes at the site and to develop a model of natural attenuation, which can be applied to other AMD sites.

Materials and Methods

Field Monitoring

A sampling network (Figure 1) was established to monitor the changes in surface and groundwater chemistry at the Davis Mine AMD site. Surface water samples were collected from the mine drainage effluent (C1 & C2) and from Davis Mine Brook both upstream (A4) and downstream (B1-B4) of the mine drainage effluent. Multi-level groundwater wells (wells 1-9) were installed within the area of exposed waste rock and along the periphery. Sampling ports were installed near the top of the water table (~1-3 m depth) and 1.5 m into bedrock. Groundwater and surface water samples were collected once a month for 13 months in 2003-2004 and included field measurements of PH, oxidation reduction potential (ORP), water table elevation, and surface water conductivity. Wells were purged of standing water prior to sampling. Groundwater samples for anion and cation analysis were filtered in-line in the field using 0.45μm filters (Millipore, Billerica, MA, USA). Samples for cation analysis were acidified with 1 drop of 6 N nitric acid in the field to prevent metal precipitation. Field samples were returned to the laboratory for cation, anion, and total organic carbon (TOC) analysis, as discussed below. Samples were also taken from selected wells for microbial community analysis.

Microbial Community Structure

Sediment cores from the initial nine well installation sites and a nearby control site were harvested under anoxic conditions and stored in a nitrogen atmosphere at 4°C in the dark. Subcores were taken at a depth of approximately 1.2 in below the surface for total DNA extraction by ballistic cell destruction. A standard DNA extraction protocol (*10*) was optimized for acidic sediments, and DNA was purified over a Sepharose column (Sepharose 4B; Sigma, St. Louis, MO, USA). DNA extraction was followed by PCR amplification, cloning of 16S rRNA marker genes into sorted clone libraries, and sequencing and detailed phylogenetic analyses, as described by Stout and Nüsslein (*11*). One hundred clones from each library were sequenced and phylogenetically analyzed. Possible chimeric sequences were detected using CHIMERA_CHECK in the Ribosomal Database Project II and also with the software Bellerophon (http://foo.maths.uq.edu.au/~huber/bellerophon) (*12*).

Microcosm Studies

Microcosms were constructed from Davis Mine sediment and groundwater with added glycerol and nutrients. Sediment and groundwater were obtained from four areas of the site (wells 1-4). An additional sample (designated as background) was taken from a nearby homeowner's yard. Aquifer material was collected from each location using a hand auger at approximately a 1 in depth. The sediment was extruded into sterile glass sample containers under anaerobic conditions. Groundwater for media preparation was collected from wells corresponding to the sediment sample locations.

Triplicate live microcosms and one killed (autoclaved) control from each of the five sites were poised at initial pH values of 2, 3, 6, and 7 (total of 100 bottles). Microcosms were constructed in 125 mL glass serum bottles with 100 mL of groundwater with added glycerol (2 mM), $(NH_4)_2SO_4$ (68 mg/L) and KH_2PO_4 (6.3 mg/L). The medium was sparged with a CO_2/N_2 (30%/70%) gas mixture for 15 minutes. The pH was adjusted by adding 18 M sulfuric acid or 10 M sodium hydroxide to the bottles as needed. Five grams of site sediment were added as inoculum to each of the bottles in an anaerobic glove box (Coy Laboratory Products Grass Lake, MI, USA). Microcosm samples were monitored for pH, ORP, anions, cations, and TOC over a six-month period.

Chemical Analysis

Principal cations (Na, K, Ca, Mg), transition elements (Fe, Cu, Mn, Zn), silicon (Si) and other AMD indicator species (Al, Pb) were measured using a Spectro M 120 inductively coupled plasma optical emissions spectrophotometer (Spectro Analytical Instruments, Kleve, Germany). Anions (Cl, SO_4, NO_2, NO_3, F, Br, PO_4) were determined using a Lachat 5000 Ion Chromatograph (Hach Company, Loveland CO, USA). TOC was measured using a Shimadzu 5000 TOC analyzer (Columbia, MD, USA).

Results and Discussion

Field Monitoring

Average pH, sulfate, and iron concentrations in shallow and bedrock groundwater for 12 months of sampling indicate that the contamination is narrowly and shallowly confined to a low-lying drainage area containing the pyritic tailings where AMD is generated (wells 3, 4, 5, and 8) (Figs. 2-4). Analytical accuracy for the field data is ± 5%, so that the contours shown are reliable indicators of the general subsurface water chemistry. Hydrogeologic data indicate AMD is directed south by inflow of groundwater from surrounding higher topographies and discharges into Davis Mine Brook near the effluent stream confluence. Contamination is greatest in shallow groundwater flowing through the tailings piles, which contains Fe > 170 mg/L, SO_4^{2-} > 1,100 mg/L and pH < 3. Zn, Cu and Pb are leached out by the acidity and are also most concentrated within the tailings piles (data not shown). Bedrock groundwater below the tailings is less acidic and lower in dissolved constituents, which may be due to dilution, silicate mineral dissolution or microbial attenuation.

Seasonal variability in sulfate concentration was minimal, except for shallow groundwater in wells located in the tailings piles (wells 3, 5, and 8), which showed great variability with a maximum sulfate concentration in the summer and fall (data not shown). Metal concentrations were most elevated when groundwater recharge occurred in autumn and spring. The elevated water table during these times may flush metal-sulfate precipitates into the pore water resulting in elevated dissolved concentrations at these times.

Natural attenuation by microbial sulfate or Fe(III) reduction is likely to occur along the periphery of the contaminant plume, where geochemical data indicate groundwater conditions are more reducing. Microbial activity would result in less acidity and lower concentrations of sulfate and iron, coupled with

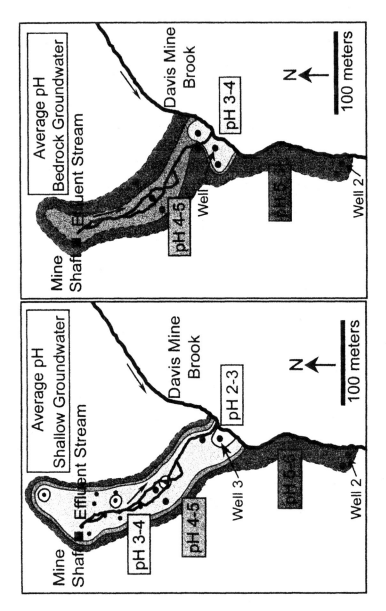

Figure 2. Average pH in shallow and bedrock groundwater. Black dots indicate well locations.

113

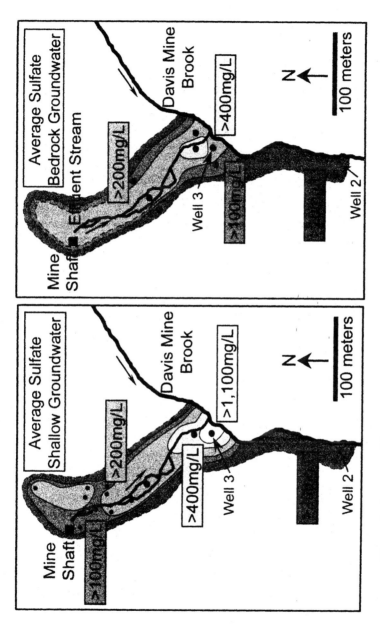

Figure 3. Average sulfate concentrations in shallow and bedrock groundwater. Black dots indicate well locations.

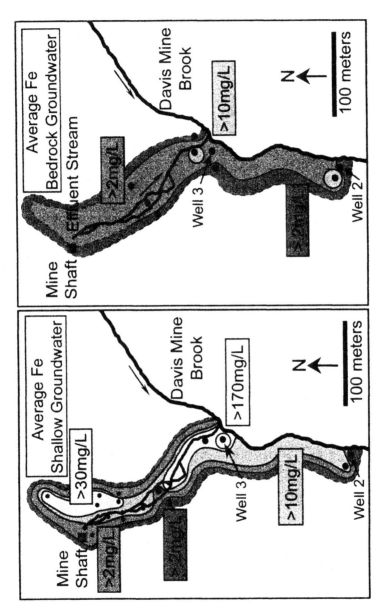

Figure 4. Average Fe (total) concentrations in shallow and bedrock groundwater. Black dots indicate well locations.

lower ORP. Zones where these conditions occur at Davis Mine include bedrock groundwater in the vicinity of wells 3, 5, and 8, which has lower acidity and ORP than the shallow tailings groundwater, where AMD is generated. There is also less sulfate and iron in bedrock groundwater in the vicinity of wells 3 and 8, although not near well 5. In addition, dissolved iron in bedrock groundwater in the vicinity of well 3 is mostly Fe(II). This suggests a zone of microbial sulfate or Fe(III) reduction near the vicinity of the tailings-bedrock interface along the bottom of the acidic drainage plume.

Less oxidizing groundwater conditions were also found in the periphery of the tailings piles near wells 1, 2, and 9. This groundwater is also less acidic, pH approximately 5.5, with lower concentrations of sulfate and iron than in other areas. These conditions indicate microbial sulfate or Fe(III) reduction may occur along the periphery of the area of exposed waste rock.

Microbial Community Analysis

Culture-independent community analyses from upstream and midstream zones exhibited high diversity within the Bacteria domain, with dominant organisms including acidophiles commonly found in AMD, such as *Thiobacillus thiooxidans, Acidobacterium capsulatum,* and *Nitrospirca multiformis.* While well 2 was dominated by close relatives to presently uncultured acidobacteria, iron-oxidizing bacteria, such as *Leptothrix ferrooxidans,* were found amongst the dominant species from the midstream zone (well 3). Phylogenetic assignments could not identify most of the clones as previously isolated species, but in the present research, close relationships were able to be established with known species. For example, 61% of the entire clone library (n= 100) prepared for well 2 could only be related to other, presently uncultured bacterial species. For well 3 about 30% had no match with previously cultivated organisms. Still, many clones could be grouped with their closest relatives based on phylogeny, and all dominant community members for wells 2 and 3 were able to be sorted into specific functional groups (Figure 5). Clones were arranged into Fe(II)-oxidizing, Fe(III)-reducing, sulfate-reducing, close relatives to sequences cloned from other AMD sites, and known acidophiles. Those that could not be placed were separated under the term "others."

When the community composition of well 3 with well 2 were contrasted, both wells were dominated by acidophiles and organisms found previously at other AMD sites. However, while well 2 exhibited close relatives to Fe(III)-reducing and sulfur-/sulfate-reducing bacteria, well 3 did not show these types of anaerobes, but a pronounced presence of iron oxidizers instead.

Finding a strong presence of a Fe(II)-oxidizing community amongst the dominant members of well 3 presented this location as an AMD-generating site,

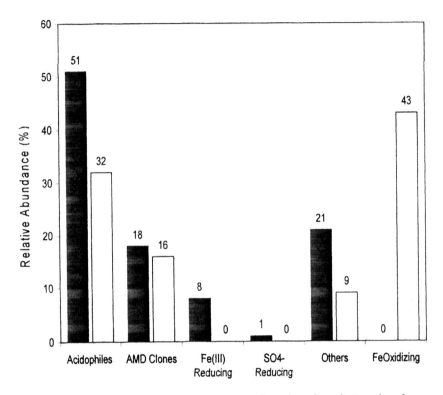

Figure 5. Microbial community composition based on the relative abundance of marker genes (16S rRNA; in percent) and their affiliation with known functional groups. Solid bars, peripheral well (2); white bars, tailings pile well (3).

while well 2 showed mostly anaerobic modes of metabolism, such as Fe(III) and sulfate reduction. Analogous findings in neighboring wells of the respective midstream and downstream areas supported a similar interpretation for both AMD-generating metabolic types of organisms associated with wells in tailings piles, and, in contrast, AMD-attenuating metabolic types associated with wells in the periphery of tailings piles (data not shown). This site-specific community composition supported rejection of the hypothesis that finding AMD-related groups of organisms is only a result of local hydrological conditions at our sampling depth. At the AMD-generating site (well 3) it was also found that the two species most widely studied and implicated for their role in AMD production, *Acidithiobacillus ferrooxidans* and *Leptospirillum ferrooxidans*. In all wells, the majority of the dominant organisms detected were newly discovered or organisms only recently associated with acid-leaching environments.

Microcosm Studies

There were no significant differences between live and killed microcosms initially poised at pH 2, suggesting that microbial activity was absent or insignificant at pH 2 (data not shown). The pH increased in microcosms initially poised at pH 3 from both AMD-generating (well 3) and -attenuating (well 2) zones to greater than 5 (Figure 6). This increase in pH and accompanying decrease in ORP (Figure 7) were not observed in killed controls, suggesting the activity of acidophilic iron-reducing bacteria capable of generating reducing conditions necessary for Fe(III) and sulfate reduction at pH as low as 3. pH and ORP were also negatively correlated in field groundwater samples from the site. Results of microcosms poised at pH 6 and 7 were similar (only pH 7 results shown). The pH of both live and killed microcosms initially set at pH 6 and 7 varied slightly over the first 100 days of incubation and then decreased slightly, possibly due to abiotic or microbial oxidation of pyrite coupled with Fe(III) reduction. ORP in pH 6 and 7 microcosms decreased over the six-month incubation period.

Total iron concentrations over time in microcosms from AMD-generating and -attenuating zones are shown in Figure 8. The greatest increases in total iron concentrations were observed in live microcosms from the AMD-generating (well 3) microcosms, an indication of Fe(III) reduction to Fe(II), which has greater solubility. Dissolved iron in bedrock groundwater from well 3 was shown to be mostly Fe(II) supporting the activity of iron reducers; however, community analysis failed to find known Fe(III)-reducing anaerobes in samples from this area.

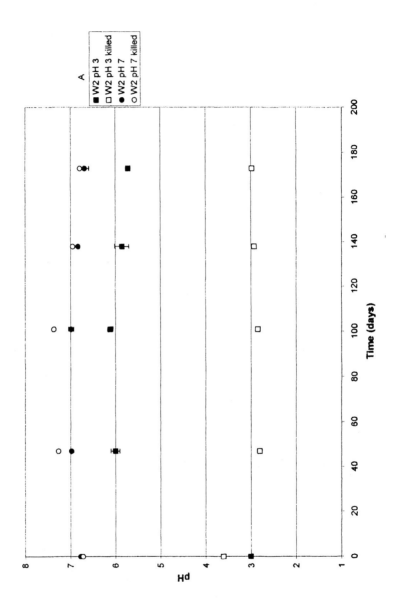

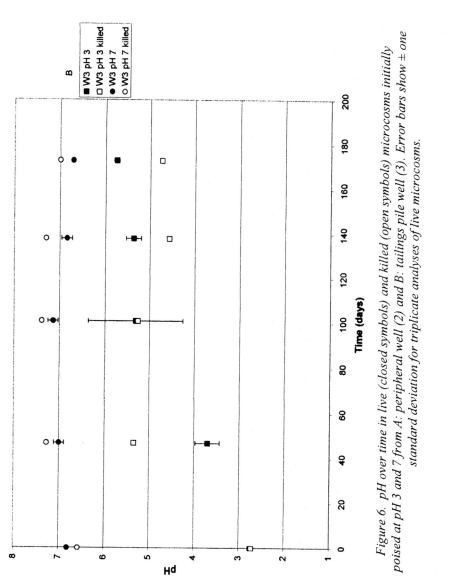

Figure 6. pH over time in live (closed symbols) and killed (open symbols) microcosms initially poised at pH 3 and 7 from A: peripheral well (2) and B: tailings pile well (3). Error bars show ± one standard deviation for triplicate analyses of live microcosms.

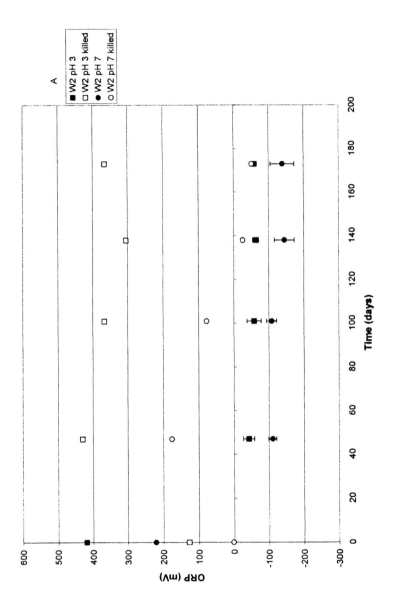

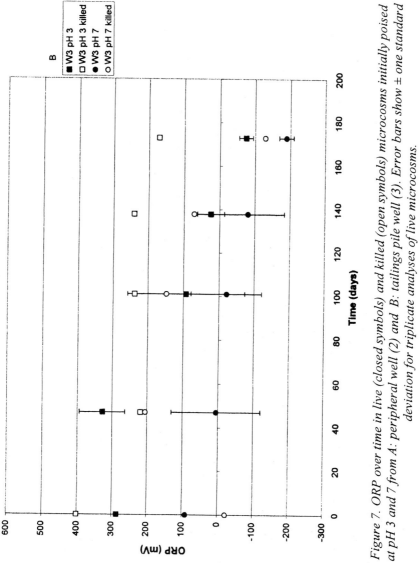

Figure 7. ORP over time in live (closed symbols) and killed (open symbols) microcosms initially poised at pH 3 and 7 from A: peripheral well (2) and B: tailings pile well (3). Error bars show ± one standard deviation for triplicate analyses of live microcosms.

122

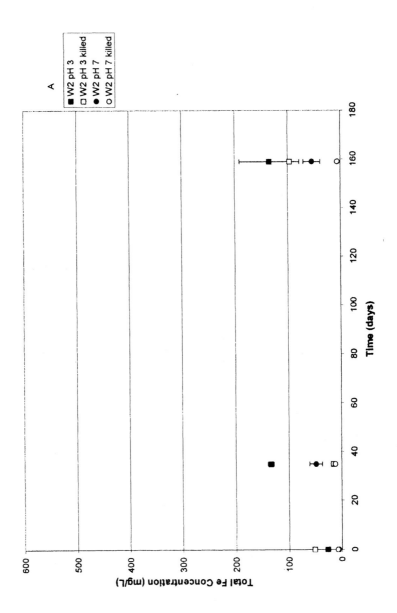

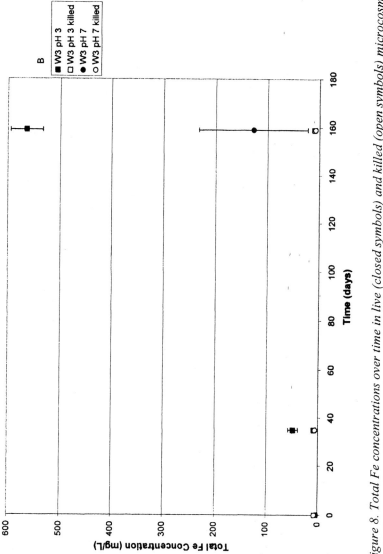

Figure 8. Total Fe concentrations over time in live (closed symbols) and killed (open symbols) microcosms initially poised at pH 3 and 7 from A: peripheral well (2) and B: tailings pile well (3). Error bars show ± one standard deviation for triplicate analyses of live microcosms.

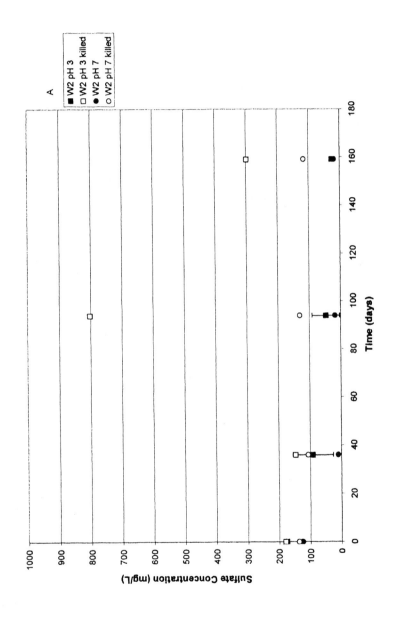

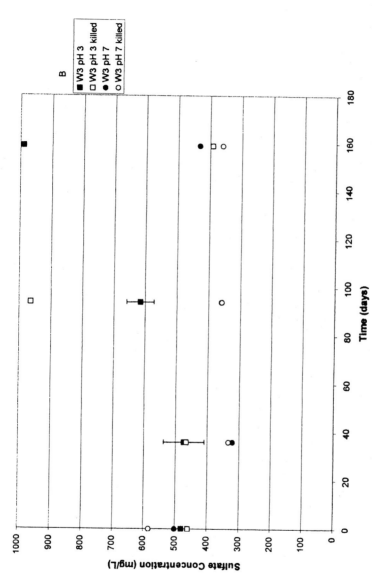

Figure 9. Sulfate concentrations over time in live (closed symbols) and killed (open symbols) microcosms initially poised at pH 3 and 7 from A: peripheral well (2) and B: tailings pile well (3). Error bars show ± one standard deviation for triplicate analyses of live microcosms.

Sulfate concentrations decreased over time in microcosms from AMD-attenuating zones and increased in microcosms from AMD-generating zones (Figure 9), suggesting that sulfate-reducing bacteria were restricted to peripheral zones with higher pH. Acidotolerent sulfate-reducing bacteria have recently been described to be able to survive in mixed cultures at pH as low as 2.8 (*13*) and to reduce sulfate after other microbial flora increase the pH of the environment to approximately 5.5.

Background samples, taken from a residence near the Davis Mine also exhibited Fe(III)- and sulfate-reducing activity, similar to results from the AMD-attenuating zone (data not shown). Although background groundwater had a pH of 6.5 and community analysis showed no signs of acidophiles, the site was a forested area within the Hawley pyrite formation, providing sulfate and organic carbon that can select for sulfate-reducing bacteria.

Sulfate concentrations increased in both the pH 3 and 7 microcosms from the AMD-generating zone, possibly due to dissolution of sulfate minerals, such as jarosite, or pyrite oxidation coupled with Fe(III) reduction. These processes can occur in the absence of microorganisms, as evidenced by the changes in sulfate concentrations in the killed controls.

Conclusions

The results of field, microbial community analysis, and microcosm studies suggested that dilution, silicate mineral dissolution and Fe(III) and sulfate reduction were the primary factors attenuating acidity leaving Davis Mine. Hydrogeologic data indicated than inflow of uncontaminated groundwater restricted the contamination to a narrow area below the tailings piles. Bedrock and peripheral groundwater was less contaminated than tailings groundwater, possibly due to the dissolution of silicate minerals. Microcosm studies indicated that microbial Fe(III) reduction was widely distributed, in both AMD-generating and -attenuating zones, while sulfate reduction was restricted to regions on the periphery of the contaminant plume. The detection of Fe(III)- and sulfate-reducing bacteria in our samples expanded the overall knowledge base pertaining to iron and sulfur cycles in AMD. The results expanded knowledge of the biodiversity of acid mine drainage environments and also had relevance to the microbiology of industrial bioleaching and to the understanding of geochemical iron and sulfur cycles.

Acknowledgments

We thank the students and teacher-scholars who have contributed to the field and laboratory effort of this project: Cristine Barreto, Liam Bevan, Jennifer Clark, Phil Dixon, Jason Jean, Mercedita Monserrate, Melissa Russell, Ashmita Sengupta, and Janice Wing. This research is being supported by the U.S. National Science Foundation, CHE-0221791.

References

1. Johnson, P.B. *Int. Biodeter. Biodegr.* **1995**, *35*, 41-58.
2. Evangelou, V.P.; Zhang, Y.L. *Crit. Rev. Environ. Ski. Technol.* 1995, *25*, 141-199.
3. Kleinmann, R.L.P.; Crerar, D.A.; Pacelli, R.R. *Mining Eng.* **1981**, *33*, 300-305.
4. Berger, J.W.; Bethke, C.M.; Krumhansl, J.L. *Appl. Geochem.* **2000**, *15*, 655-666.
5. Strömberg, B.; Banwart, S.A. *Appl. Geochem.* **1999**, *14*, 1-16.
6. Widdel, F. In *Biology of Anaerobic Microorganisms;* Zehnder, A.J.B., Ed.; Wiley: New York, NY, 1988; pp 469-585.
7. Baker, B.J.; Banfield, J.F. *FEMS Microbiol. Ecol.* 2003, *44*, 139-152.
8. Cummings, D.E.; March, A.W.; Bostick, B; Spring, S.; Caccavo, F. Jr.; Fendorf, S.; Rosenzweig, R.F. *Appl. Environ. Microbiol.* **2000**, *66*, 154-162.
9. Gál, N. E. PhD Thesis, University of Massachusetts, Amherst, MA, **2000**.
10. *Short Protocols in Molecular Biology;* Ausubel, F.M..; Brent, R.; Kingston, R.E.; Moore, P.D.; Seidman, J.G.; Smith, J.A.; Struhl, K., Eds.; 3rd edition; John Wiley & Sons: New York, 1995.
11. Stout, L.M.; Nüsslein, K. *Appl. Environ. Microbiol.* **2005**, *71*, 2399-2407.
12. Huber, T.; Faulkner, G.; Hugenholtz, P. *Bioinformatics* **2004**, *20*, 2317-2319
13. Küsel, K; Roth, U.; Trinkwalter, T.; Peiffer, S. *Microbial Ecol.* 2001, *40*, 238-249.

Chapter 8

Molecular Methods Used for Evaluating Subsurface Remediation

Andria M. Costello

Department of Civil and Environmental Engineering, Syracuse University, Syracuse, NY 13244

Advances in the field of molecular biology have revolutionized the tools available to study microorganisms in the environment. Recently developed molecular biological techniques are providing insight into the diversity, community structure, and function of environmentally relevant microbes. While providing fundamental knowledge regarding the metabolic capabilities of microorganisms, the information being provided using the new methods is critical for addressing issues regarding the implementation of bioremediation for subsurface remediation.

Bioremediation is a technology that harnesses the metabolic potential of microorganisms to clean up contaminated environments (*1*). As a technology, bioremediation is continually evolving and expanding, especially in response to the growing need to maintain clean soil, air, and water (*2*). In many processes utilizing bioremediation for clean-up activities, the "bio" or microbial component has often been treated as a "black box." Historically, as long as remediation was taking place, little to no effort was expended in defining the microbial communities responsible for contaminant degradation. Within the last two decades, however, the advancement of environmental microbiology and biotechnology has changed our understanding of the role of microbial communities in the environment (*3*). Specifically, the development and implementation of molecular biological tools for use in bioremediation has facilitated the study of microorganisms in their natural environment and has paved the way towards a more complete understanding of the metabolic capabilities of in situ microorganisms.

It is currently estimated that less than 1% of the 10^9 bacterial cells in a gram of soil are capable of being cultured in the laboratory (*4*). However, much of the information that has been obtained in bioremediation studies has utilized data gathered from laboratory studies of cultured microorganisms. These data were then extrapolated to what might be expected in field studies where organisms similar to the laboratory cultures would be likely to be found. In these cases, the activities of the cultures in the laboratory were hypothesized to reflect those activities that w ould be expected to occur in s itu. In the absence of any pure cultures or laboratory data, in situ microorganisms were considered an indiscreet consortium, and the metabolic potential was correlated to observed breakdown products of the contaminant of interest or other physical or chemical indicators. The introduction of molecular biological tools into bioremediation research has enabled researchers to begin to link the diversity of environmental microorganisms with their community structure and degradative capabilities, all without the need for cultivation.

Recently, molecular biology based methods have been used to monitor the performance of in situ bioremediation and to diagnose the potential for successful bioremediation before deploying a field system (*5*). For bioremediation to be effective, it is necessary to have a means of identifying and quantifying both the organisms that are responsible for contaminant removal and their activities (*6*). Molecular biological methods are valuable diagnostic tools for these purposes and can help to demonstrate that contaminant loss in the field corresponds to biological removal processes and is not simply due to the contaminant being transferred from one location to another (*5*). Applying new molecular biological methods to bioremediation can provide important information that cannot be obtained by other microbiological analyses. The use of many of the available molecular techniques has become routine for analyzing microbial communities involved in bioremediation. These tools are enabling valuable insights into the relationship between diversity, structure, and function in microbial communities that are directly relevant to environmental biotechnology applications (*3*). This article discusses recent and evolving uses of molecular biological tools for the study of microbial communities involved in subsurface remediation.

DNA Extraction

The use of molecular methods for analyzing environmental microbial communities relies on the extraction of nucleic acids of high quality and quantity for use in subsequent analyses. However, direct extraction of nucleic acids from soils, sediments, sludge, and water usually results in the coextraction of contaminating materials such as humic and fulvic acids, the breakdown products of organic materials (*7*). Furthermore, in bioremediation studies of polluted environments, the presence of contaminating compounds and/or heavy metals could interfere both with the extraction of the nucleic acids as well as subsequent molecular biological assays (*7*).

Historically, two different techniques have been used for isolating DNA from environmental samples: the direct lysis method and indirect cell extraction and lysis method. The direct method relies on the in situ lysis of microorganisms in environmental samples and subsequent recovery and purification of the DNA (*8-10*). One of the advantages of the direct method is that it tends to recover DNA from organisms that are strongly sorbed to environmental samples (*11*). However, the direct method also usually involves some type of mechanical disruption of the cells, which can lead to undesirable shearing of the DNA. Furthermore, humic acids are often coextracted when using the direct method and v arying pH o f e nvironmental s amples c an i nterfere w ith D NA e xtraction (*11*).

To circumvent some of the problems associated with the direct lysis method, indirect methods were developed where the extraction of microbial cells from environmental samples precedes cell lysis and DNA purification, in general, this indirect approach produces longer DNA fragments (which can be useful in some molecular techniques such as metagenomic analyses) and contamination of coextracted inhibitory compounds is low (*12*). While both methods strive to produce the highest quality and quantity of DNA for use in subsequent molecular analyses, the direct lysis method has been the most widely implemented technique for DNA isolation from environmental samples. Many commercial DNA extraction kits have become available in recent years for use with a variety of environmental sample types. These kits utilize the direct lysis method a nd o ffer e ase o f u se, high t hroughput, a nd t he p otential for reduced variability of extraction between samples and users (*13*).

Polymerase Chain Reaction

The polymerase chain reaction (PCR) was an idea conceived in the early 1980s by Kary Mullis, who, at the time, worked for California biotechnology company Cetus (bought out by Chiron). Now in routine use in all areas of medicine, science, and biotechnology, PCR has become one of the most common techniques in molecular biology, and many subsequent molecular techniques rely on PCR as a first step for generating sufficient quantities of nucleic acids. PCR is a procedure that generates large quantities of a specific DNA sequence using a three-step cycling process (*14*). It requires two oligonucleotide primers flanking the gene of interest, template DNA, a thermostable DNA polymerase, and the four deoxyribonucleotides (*14*). A typical process involves denaturation of the template DNA, renaturation and binding of the oligonucleotide primers, and synthesis of new DNA strands via *Taq* DNA polymerase. The entire process occurs in an automated, programmable thermal cycler, which are available from field-portable to industrial-throughput capacity. The end product of a PCR reaction is millions of copies of a specific DNA sequence, and this is advantageous for the ultimate study of microorganisms and their activities in the natural environment.

Real-time PCR

Conventional PCR yields no quantitative measurement of the number of copies of DNA in the original sample. However, real-time PCR has the capability to quantify the number of gene copies present in the initial sample (*15*). In real-time PCR, fluorescence is incorporated into DNA during amplification, and the amount of fluorescence is used to quantify the number of PCR products at the end of each three-step cycle. Currently, fluorescent technologies include SYBR-Green, TaqMan, and molecular beacons (*15*). Real-time PCR has seen increasing use in the study of subsurface remediation because it allows researchers to gain a quantitative view of the genes responsible for degradation in a particular environment. Beller et al. assessed anaerobic, toluene-degrading bacteria in aquifer sediment contaminated with hydrocarbons (*16*). Hydrocarbon-degrading bacteria were targeted in real-time PCR by using primers for the *bssA* gene, a catabolic gene associated with the first step in anaerobic toluene and xylene degradation (*16*). In microcosms, the authors showed that the *bssA* gene increased in number 100- to 1 000-fold during the period of maximum toluene degradation and that the microcosms with the most *bssA* genes showed the most toluene degradation, indicating a role for this gene in toluene removal in situ (*16*). Ritalahti and Löffler used real-time PCR to detect *Dehalococcoides* populations in microcosms from pristine and 1,2-dichloropropane-contaminated sediments using the 16S rRNA gene (*17*). They determined that false-negative results were obtained with direct PCR and that greater than 10^4 cells per ml of culture was required to yield results, while real-time PCR could detect *Dehalococcoides* genes present as low as 10^3 cells per ml (*17*). Ritalahti and Löffler used real-time PCR to quantitate the increase in *Dehalococcoides* in the presence of sufficient chlorinated electron acceptor and suggested the use of this method in conjunction with other molecular techniques for identification of these organisms in mixed degrading cultures (*17*).

Reverse Transcriptase PCR

While traditional PCR has typically been used to amplify copies of DNA sequences, these sequences indicate that a particular gene is present in an environmental sample but do not indicate anything about the activity of a gene in situ. It is often desirable to know about the activity of particular genes in the environment and this can be accomplished using reverse transcriptase PCR (RT-PCR). In RT-PCR, mRNA is extracted from environmental samples, and cDNA is made using the enzyme reverse transcriptase. The cDNA is used as template in subsequent PCR reactions to generate copies of a desired gene. One of the challenges in using RT-PCR for environmental samples is that mRNA has a short half-life and is labile to RNases present in the samples (*15*). Thus, the detection limits for mRNA are not as low as for DNA because of instability and degradation (*18*). Because of these limitations, RT-PCR has not been used extensively with environmental samples to date. In one of the first applications of RT-PCR to study subsurface bioremediation, Holmes et al. studied the *nifD* gene, one of the dinitrogenase genes, in *Geobacteraceae* in sediments contaminated with crude oil (*19*). mRNA was extracted from chemostats and sediments and assessed for the presence of the *nifD* gene (*19*). From their studies, the authors concluded that *Geobacteraceae* in subsurface sediments or in petroleum-contaminated aquifers express *nifD* to stimulate dissimilatory metal reduction (*19*). Thus the authors were able to gather information regarding the metabolic status of *Geobacteraceae* during in situ bioremediation.

Restriction Fragment Length Polymorphism

Restriction fragment length polymorphism (RFLP) is a molecular method that utilizes DNA generated via PCR. In RFLP of environmental samples, PCR-amplified DNA is cloned, and the clones are subjected to restriction enzyme digestion by tetrameric restriction enzymes (Figure 1). This process generates a genetic fingerprint of the clone due to the fact that different species will have differing fragment lengths because of variations in their gene sequences (*20*). The resulting restriction fragment length polymorphisms can be visualized on low-melting agarose (or polyacrylamide) gels (Figure 1), and the different banding patterns can be used to screen clones and provide a measure of community diversity (*21*). In studies of environmental samples, RFLP has been

134

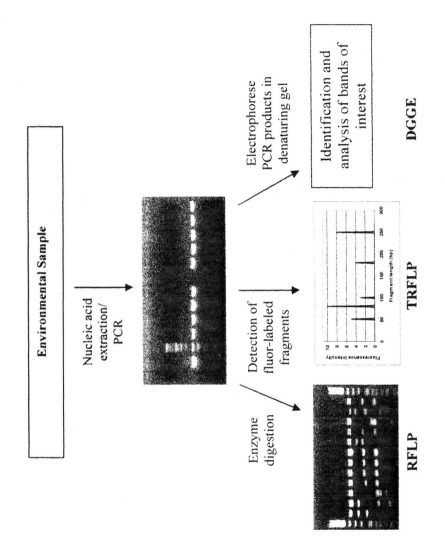

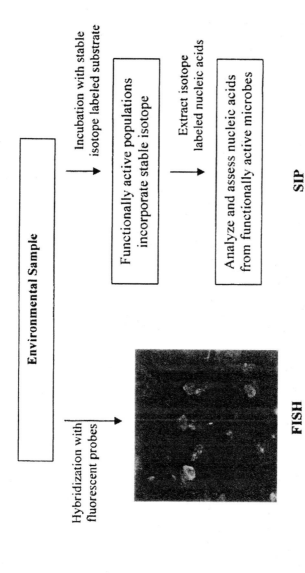

Environmental Sample

Incubation with stable isotope labeled substrate

Functionally active populations incorporate stable isotope

Extract isotope labeled nucleic acids

Analyze and assess nucleic acids from functionally active microbes

SIP

Hybridization with fluorescent probes

FISH

Figure 1. Use of molecular tools for analyzing microbial populations in environmental samples.

used primarily to screen clone libraries of 16S rRNA genes amplified from total extracted DNA. This type of RFLP is commonly referred to as amplified ribosomal DNA restriction analysis (ARDRA) (*21*).

RFLP of microbial communities involved in subsurface remediation has been extensively described in the literature for a number of different contaminants. In their work in an aquifer contaminated with jet fuel and chlorinated solvents, Dojka et al. assessed the diversity of microorganisms involved in intrinsic remediation at three distinct redox zones within the aquifer (*22*). After RFLP analysis of over 800 clones, the authors were able to identify the major microorganisms in each redox zone and developed a hypothesis about carbon flow and hydrocarbon degradation in the aquifer (*22*). Hou and Dutta utilized operational taxonomic units (OTUs), common restriction fragments used to distinguish between taxonomic groups, to characterize the diversity of 16S rRNA genes from polychlorinated biphenyl (PCB)-contaminated sediments (*23*). Using DNA extracted from a *para-* and *meta*-dechlorinating consortium, they identified 9 different *Clostridium* species capable of PCB degradation (*23*). Because various *Clostridium* species were present in the clone banks, albeit in varying amounts, the authors concluded that the *para-* and *meta*-degrading species are phylogenetically different and both are necessary in consortia for complete PCB degradation (*23*).

While ARDRA is one of the common techniques used in assessing the diversity of environmental microbial communities, RFLP analysis of functional genes is also used. However, the database of sequence information regarding various functional genes is not as developed as for the 16S rRNA genes. In their work with chlorobenzoate-degrading microbial communities, Peel and Wyndham were able to enrich isolates that contained genes for the chlorocatechol 1,2-dioxygenase, the 3-chlorobenzoate 3,3-(4,5)-dioxygenase, and the 4-chlorobenzoate dehalogenase from contaminated groundwater and sediments (*24*). Using RFLP, they showed that the 3-chlorobenzoate 3,3-(4,5)-dioxygenase genes were conserved in all of the degrading isolates while the chlorocatechol 1,2-dioxygenase and 4-chlorobenzoate dehalogenase genes varied among different isolates (*24*). None of the PCB-degrading genes were found in isolates from uncontaminated samples. The results from the RFLP study allowed the authors to make hypotheses regarding the role of gene mobilization and horizontal gene transfer in organisms responsible for biodegradation of chlorobenzoate (*24*).

Although RFLP has provided insight into previously uncharacterized environments, its use is limited to information regarding diversity of microbial communities. Furthermore, in diverse communities the usefulness of RFLP becomes limited because large numbers of banding patterns are hard to distinguish during visualization on gels. Nevertheless, RFLP was one of the first molecular biological analysis tools to be used in characterizing subsurface microbial communities and its use is continued today in both basic and applied research.

Terminal Restriction Fragment Length Polymorphism

Terminal restriction fragment length polymorphism (T-RFLP) is another effective method for studying the diversity of microbial communities. T-RFLP follows the same principles of RFLP except that a fluorescently-labeled primer (or primers) is incorporated during the PCR amplification step. After restriction digestion of the PCR products, the fluorescently-labeled terminal fragments are analyzed by an automated DNA sequencer, and the sizes of the fragments are digitally recorded in electropherograms (Figure 1) (25). Because only the terminal fragments are detected in T-RFLP, the banding patterns are greatly simplified from traditional RFLP and each visible band theoretically represents one OTU, or ribotype, that can be used to determine the genetic diversity of the community (Figure 1) (21). Although T-RFLP represents advantages over RFLP, it can only analyze fragments from ~50-500 bp, and it is possible that several ribotypes may have the same terminal restriction fragment (20). In spite of these potential limitations, T-RFLP has rapidly overtaken RFLP as the preferred molecular method to study the diversity of subsurface microbial communities. To aid researchers in T-RFLP analyses, the Ribosomal Database Project (RDP) maintains the T-RFLP Analysis Program (TAP)-TRFLP program, which aids in selecting the appropriate primer and enzyme combinations for analysis of 16S rRNA genes (26, 27).

T-RFLP studies have been carried out in a number of contaminated environments. Petrie et al. studied the diversity of iron(III)-reducing microbial communities from uranium-contaminated sediments and compared the T-RFLP results to sequences obtained from 165 rRNA clones (28). The results from the T-RFLP and cloning suggested very little diversity in enrichment cultures amended with lactate or acetate (28). Interestingly, the contaminated enrichments amended with lactate or acetate contained fragments most closely associated with *Anaeromyxobacter* when analyzed by T-RFLP but only *Geobacter*-type sequences were found in background enrichments amended with lactate (28). This was the first report of *Anaeromyxobacter* being detected in the subsurface and studied using cultivation-independent techniques (28). In a trichloroethene (TCE)-contaminated deep, fractured basalt aquifer, the natural groundwater microbial community was compared to an enrichment culture of the indigenous microorganisms (29). While the mixed enrichment culture completely dechlorinated TCE to ethene, T-RFLP studies showed that it was less diverse than the natural microbial community (29). The laboratory enrichment culture contained 22 unique phylotypes, compared to 43 in the indigenous community, indicating that although degradative function was maintained, almost half of the original microorganisms did not survive during laboratory incubation (29). Another study used T-RFLP to show that the bacterial community involved in in situ remediation of petroleum waste sludge in landfarming sites decreased over the 14 month study period although the observed degradative capabilities remained the same (30). Thus, it is clear that T-RFLP can provide insight into community dynamics that was previously not available using other molecular techniques.

Denaturing Gradient Gel Electrophoresis

Denaturing gradient gel electrophoresis (DGGE), an approach for analyzing the diversity of complex microbial communities, was described first in the early 1990s (31). In DGGE, DNA extracted from environmental samples is used as template in PCR with primers targeting a desired gene (for example, the 16S rRNA gene). The 5' end of one of the primers contains a ~40-bp, GC-rich sequence, called a clamp, which prevents formation of single-stranded DNA (ssDNA) (20). The amplified PCR products are separated into discrete bands during polyacrylamide gel electrophoresis containing a linearly increasing gradient of DNA denaturant (urea and formamide). DGGE is based on the principle that sequence differences in gene fragments will create a difference in the melting behavior of the double-stranded DNA so that individual fragments will migrate through the gel to stop at a unique position for each gene with a different sequence. Thus, the diversity of a microbial community can be estimated based on the banding patterns produced in the DGGE gel (32). DGGE bands of interest can be excised from the gel and sequenced, allowing identification of the community members represented by individual bands. Temperature gradient gel electrophoresis (TGGE) acts upon the same principles except that the gradient is temperature instead of chemical denaturants (21).

The use of DGGE in subsurface remediation studies has focused primarily on the analysis of community dynamics using the 16S rRNA gene. Baker et al. studied a TCE-contaminated aquifer before and after the addition of methane, oxygen, nutrients, and air to stimulate the growth of methane-oxidizing bacteria (33). Using DGGE, they analyzed the microbial communities and were able to show that biostimulation with methane caused a change in the dominant microorganisms in the aquifer from Enterobacteria and *Pseudomonas* spp. to type I methane oxidizers (33). Because type I methane oxidizers are known to cometabolize TCE poorly, the authors were able to relate their DGGE results to the failure of the biostimulation procedure in the aquifer (33). In a study of the degradation of the insecticide chlorpyrifos in soils, Singh et al. used DGGE to investigate the transfer of degradative capabilities from a soil in Australia to one in the United Kingdom (*34*). They were able to show that they had isolated the chlorpyrifos-degrading organism from the Australian soil by matching the DGGE band of the isolate with an identical one from the soil profile (*34*). This was important because it allowed the researchers to conclusively link an isolate with observed degradative capabilities. Furthermore, Yang et al. used DGGE to compare the community compositions of *Dehalococcoides* and *Desulfitobacterium* involved in reductive dehalogenation of chlorinated ethenes from a tetrachloroethene (PCE)-contaminated aquifer (*35*). They followed enrichment cultures and sequenced the DGGE bands that constituted major compositional changes over time (*35*). The authors were able to conclude that different enrichment conditions selected for the presence of specific groups of *Dehalococcoides* and/or *Desulfitobacterium* and that these enrichments exhibited varying degrees of dehalogenation (*35*). DGGE of 165 rRNA genes has also been w idely employed for studying the biodegradation of petroleum hydrocarbons in field studies (for example, see (*36-38*)).

While the studies of 165 rRNA genes using DGGE have provided researchers w ith i nformation o n t he c hange i n microbial c ommunity s tructure during contaminant degradation processes, it is important to understand the function of the microorganisms, as well as their diversity, to successfully implement subsurface bioremediation projects. To date, however, examples of DGGE using catabolic genes are scarce, especially as related to remediation. The most widely cited use of DGGE with catabolic genes are for the *pmoA* and *mxaF* genes, involved in methane and methanol oxidation, respectively (*39-41*). Thus, the expansion of DGGE to catabolic genes is the next logical step for this technique, especially in regards to the study of subsurface remediation.

Fluorescence In Situ Hybridization

Fluorescence in situ hybridization (FISH) is a rapid and specific method for the phylogenetic detection of microorganisms in their natural environment. FISH utilizes hybridization of a fluorescence-labeled oligonucleotide probe to intracellular rRNA in microbial cells. The labeled cells can then be detected using a number of techniques i ncluding epifluorescence microscopy, c onfocal laser scanning microscopy, microautoradiography, and flow cytometry (Figure 1) (*32, 42*). The rRNA genes are the primary targets of FISH because of their ubiquity in microbial cells and because they contain both variable and highly conserved sequences, which are important for the design of the hybridizing probes (*43*). Most studies have focused on the 165 rRNA genes due to the large databases of 16S rRNA gene sequences. The power of FISH lies in the ability to visualize the community structure and spatial arrangement of in situ microbial populations in a quantitative manner.

FISH has been employed for almost a decade in the analysis of the microorganisms involved in activated sludge, anaerobic digestors, and biofilms (*20, 43, 44*). More recently, studies of the microbes involved in subsurface remediation have been undertaken but these have lagged other studies due to the complex nature of contaminated environments. Watanabe et al. characterized the bacterial populations in petroleum-contaminated groundwater (*45*). Using 16S rRNA gene studies and DGGE, they identified two major clusters of bacteria present in the groundwater (*45*). FISH was used to further investigate one of the clusters, and it was found that microorganisms in this cluster represented up to 24% of the total microbial population (*45*). Christensen et al. examined the removal of polycyclic aromatic hydrocarbons (PAHs) from sewage sludge using FISH (*46*). They determined that enrichments provided PAHs consisted of a mixture of *Bacteria* and *Archaea,* and they used FISH to identify the *Archaea* as *Methanobacteriales* using an order-specific probe (*46*). The overall results from the study led the authors to conclude that the microbial community responsible for PAH degradation was a syntrophic culture with the bacteria oxidizing the naphthalene and the *Methanobacteriales* converting the produced hydrogen to methane (*46*). Recently, Yang and Zeyer developed novel oligonucleotide probes targeting the 16S rRNA of *Dehalococcoides* species for use in FISH (*47*). Aulenta et al. utilized the probes for the detection of *Dehalococcoides* spp. in a PCE-degrading bioreactor (*48*). The authors were able to obtain a quantitative estimate of the number of *Dehalococcoides* spp. in the bioreactor by combining the FISH results with kinetic batch tests at non-limiting hydrogen and PCE concentrations (*48*). Further method development is needed with respect to the use of FISH for analyzing the microorganisms involved in subsurface remediation.

Stable Isotope Probing

Although a significant amount of information has been gathered on the role of environmental microorganisms in subsurface remediation through laboratory, microcosm, and molecular studies, it is important to begin to study microorganisms in the complex environments in which they are normally found (*49*). One approach that is being used to identify metabolically active microbial populations in situ is stable isotope probing (SIP). Stable isotope probing is a culture-independent technique used to study the active organisms in environmental samples (*50-52*). SIP seeks to link the functional activity of a particular environment with the microorganisms responsible (*51*). The method has recently been comprehensively reviewed in an article by Dumont and Murrell (*53*).

SIP utilizes substrates enriched with stable isotopes during incubations of environmental samples (Figure 1). In their pioneering work, Radajewski et al. used $^{13}CH_3OH$ in incubations of oak forest soils and showed that ^{13}C-DNA was produced during growth of indigenous microorganisms on the substrate (54). Because the incorporation of ^{13}C into DNA enhances the density differences between labeled and unlabeled fractions, the ^{13}C-DNA can be separated from ^{12}C-DNA by density-gradient centrifugation (51, 54). The heaviest DNA (^{13}C-DNA) fraction contains the combined genomes of the microbial population that has incorporated the labeled substrate into their nucleic acids and hence represents the metabolically active microorganisms (51). In order to be able to separate the heavy (labeled) and light (unlabeled) DNA using density-gradient centrifugation, substrates must be chosen that have a large difference between the natural abundance of the heavy and light stable isotopes, and it must be possible to access compounds that are highly enriched with a rare stable isotope (51). To date, most SIP studies have utilized ^{13}C-labeled substrates, although isotopes of H, N, and O are potential candidates because of the availability of both heavy and light stable isotopes (51).

There are a growing number of examples of the use of SIP for studying microorganisms involved in subsurface remediation. Singleton et al. examined the microbial communities involved in degradation of ^{13}C-labeled salicylate, naphthalene, and phenanthrene in a bench-scale aerobic bioreactor used to treat soil contaminated with PAHs (55). SIP was followed with analysis of the microbial communities by DGGE and the construction of 16S rDNA clone libraries. The authors found that the salicylate- and naphthalene-degrading communities were dominated by the sequences belonging to the *Pseudomonas* and *Ralstonia* genera, but that the phenanthrene-degrading sequences were markedly different and belonged to the *Acidovorax* genus (55). Jeon et al. released ^{13}C-naphthalene into a shallow, unconfined aquifer contaminated with aromatic hydrocarbons (56). After extraction of DNA from SIP-enriched sediment samples, the authors were able to identify the naphthalene degraders using 16S rDNA clone libraries and T-RFLP. They were also able to match the profile of the dominant clone to a previously isolated bacterium that was most closely related to *Polaromonas vacuolata,* an organism not formerly implicated as important in naphthalene degradation (56). In addition to the phylogenetic data, Jeon et al. were able to identify identical naphthalene dioxygenase genes in both the isolate and the contaminated sediments (56).

In addition to using SIP with DNA, Manefield et al. have reported RNA-SIP and have used it to identify an organism from the *Thauera* genus that was responsible for phenol degradation in an aerobic bioreactor (57). Although RNA is potentially a more responsive biomarker than DNA because it is synthesized more rapidly in metabolizing cells, it is also more difficult to isolate from environmental samples, and studies using RNA-SIP have lagged those utilizing

DNA (57). In a recent study, Mahmood et al. examined the degradation of pentachlorophenol (PCP) in a grassland soil (58). ^{13}C-labeled RNA was subjected to DGGE and the authors found that the bacterial communities showed significant changes over an incubation of 63 days (58). Sequence analysis of selected DGGE bands showed homologies to uncultured organisms that were known hydrocarbon degraders (58).

Because SIP enriches the functionally active microbial populations, this molecular tool has begun to link studies of diversity to metabolic function. Combined with community analysis methods, such as construction of clone libraries, DGGE, and T-RFLP, SIP is providing clues about the microorganisms that are metabolically active in the environments in which they are found. Further d evelopment of this t echnique is likely to provide researchers with a window into the complex dynamics of the community structure and function of the microbial subsurface.

The Future of Molecular Methods

Array Technology

Array technology is a recently developed technique that is gaining use in studies of environmental microorganisms. Array technology can be described as any "high-through-put methodology permitting analysis of hundreds or thousands of genetic sequences in parallel" (59). The most common types of arrays are those that attach oligonucleotides or DNA fragments to glass slides or nylon membranes. In this technology, the probe or known sequence is arrayed on the solid support, and the unknown sequences are labeled and hybridized to the array (59). Array technology has been slow to be used to study microbial communities in the environment due to a number of problems with specificity, sensitivity, and quantitation (59, 60).

Nevertheless, due to its promise for providing high-throughput analysis of environmental diversity and function, studies using environmental samples are beginning to be reported. In one of the best examples of the application of this technology to study subsurface r emediation, R hee e t al. c onstructed a 50-mer oligonucleotide microarray containing most of the 2,402 known genes and pathways involved in biodegradation and metal resistance (60). They used the array to study a naphthalene-degrading enrichment and soil microcosms. Interestingly, they observed that the naphthalene-degrading genes from *Pseudomonas* were n ot detected, even t hough t hese o rganisms are c ommonly implicated and studied in naphthalene degradation (60). The authors confirmed the results from the microarray analyses with real-time PCR and were able to conclude that their array was useful for describing differences in microbial communities involved in degradation of PAHs and BTEX compounds (*60*). Although it was one of the first studies involving microarrays and biodegradation, a burgeoning of these types of studies is expected as microarray facilities become more prevalent and the technology overcomes some of the barriers to working with contaminated environmental samples.

Metagenomics

The field of metagenomics, the study of the genomes of uncultured organisms, has significantly enhanced our understanding of uncultured microorganisms (*61*). The method is rapidly finding its way into studies of environmental samples where isolates have been difficult to procure. Researchers have created libraries from soils and seawaters and obtained results that h ave g reatly i ncreased o ur k nowledge o f m icrobial c ommunities i n t hese environments (*62, 63*). One of the main advantages of using metagenomics for environmental samples is that no prior knowledge of gene sequences is needed, allowing researchers to theoretically access genes of any sequence or function (*61*). It has been asserted that metagenomics will enable researchers to construct the genomes of uncultured microorganism by identifying overlapping fragments in metagenomic libraries and sequencing the fragments to assemble each chromosome (*64*).

As reviewed in Schloss and Handelsman, there are two approaches to the analysis of metagenomic libraries: the function-driven and sequence-driven analysis (*64*). In the function-driven analysis, constructed clones are analyzed for a desired trait, and those expressing the trait are selected for subsequent sequence and biochemical analyses. In the sequence-driven analysis, metagenomic libraries are mined for genes using previously designed PCR primers and hybridization probes. Obviously, this approach relies on sequence homology between known and unknown genes and may introduce bias towards genes that are similar in sequence to those already identified.

For subsurface remediation, identification of clones expressing a degradative function could lead to the discovery of novel genes for metabolizing recalcitrant compounds. Furthermore, if phylogenetic genes were in close proximity to the degradative genes on the chromosome, it would be possible to identify the organism responsible for degradation. It is also possible that metagenomics will yield information on the metabolic capabilities of mixed microbial communities, where it has previously been difficult to tease out the relationship between the communities and the degradative functions. In combination with other techniques that enrich for a group of organisms that carry out a given metabolic function, such as SIP, metagenomics has the potential t o a dvance o ur understanding of t he i dentification o f o rganisms a nd genes that are truly responsible for functions observed in the subsurface.

Laboratory on a Chip

The interest and funding opportunities in the field of nanotechnology have skyrocketed in the last five years. One of the results of growth in this field has been laboratory-on-a-chip (LOC) devices, also known as "micro total analysis systems (μTAS)" or "biological microelectronic mechanical systems (BioMEMS)" (65). While these devices have seen implementation in various medical and biotechnology applications, little segue into the field of environmental m icrobiology h as b een made t o date (65). LOC d evices utilize microfluidics to create miniaturized versions of laboratory reactions in the space the size of a microscope slide or smaller. Some advantages of LOC devices are that only small samples of a system of interest are required for analysis and the devices are designed to be high-throughput and automated. Liu and Zhu envision t hat i t i s o nly a m atter o f t ime b efore r esearchers a re a ble t o u tilize LOC devices to carry out molecular techniques, such as PCR, T-RFLP, and construction of clone libraries, for environmental samples (65). Thus, LOC technology has the potential to revolutionize the way that we use molecular techniques to study the organisms involved in subsurface remediation.

Proteomics and Metabolomics

Proteomics and metabolomics are emerging approaches for characterizing microorganisms and the functions they carry out. Both techniques rely heavily on t he use o f various m olecular t echniques and attempt t o d escribe m icrobial function at the level of the protein and metabolism, respectively. Proteomics is the large-scale analysis of proteins and is concentrated on determining the "structure, expression, localization, biochemical activity, interactions, and cellular roles of as many proteins as possible" (66). Proteomics has been driven by the realization that information about the genes of an organism does not always e xplain t he f unction o f t hat o rganism (67). T hus, p roteomics s eeks t o link individual proteins with their specific function so that meaningful characterizations of microbial activity can be established. Metabolomics involves the analysis of changes in cell metabolites that might occur in cells in response to environmental or cellular changes (68). The objective in metabolomics is to identify metabolite differences between cells exposed to different conditions as a way to predict the metabolism of the cells (69). Metabolomics is concentrated on studying the properties of an organism's metabolism as a system and identifying metabolite-dependent regulatory interactions and their relationship to environmental conditions (*68*). Both proteomics and metabolomics are likely to play a large role in the future of subsurface remediation by advancing our knowledge of the fundamental relationship between microorganisms and the functions they are capable of carrying out under a wide range of conditions.

Conclusions

Microorganisms are responsible for the processes driving degradation, thus it is necessary to gather sufficient evidence that microbial degradative processes are possible, probable, ongoing, and sustainable in the subsurface. There now exists a wealth of molecular techniques that can be utilized to study environmental microorganisms and their role in subsurface remediation. Molecular methods are able to improve our understanding of the microbial community structure, diversity, and function in the subsurface and provide insight into the connections between the contaminant and the bioremediation process.

In many ways, the available molecular biological tools and techniques have built upon themselves and newly developed methods would not be possible without the groundwork laid by their predecessors. In other ways, the development of brand new methodologies has given rise to techniques that could not have been envisioned ten or even five years ago. Many people would agree with Head and Bailey in the belief that molecular tools may be the key to understanding and sustainably using microorganisms in environmental biotechnology applications (*3*).

The use of molecular methods for studying subsurface remediation is a merging of science and engineering practice, and the knowledge that we gain through use of these techniques is critical for addressing the prolonged issues regarding the feasibility of implementing bioremediation for removal of environmental pollutants. The future is likely to bring advances in molecular methods that cannot even be imagined, and the study of subsurface remediation can only benefit as a result.

Acknowledgements

The author would like to thank Ines Otz, Syracuse University, for her help in compiling references for this article. The author would also like to thank Mark Bremer and Jennifer Bryz-Gornia, Syracuse University, for providing data for

Figure 1. Research activities by AMC have been supported by the National Science Foundation (BES-0093513).

References

1. Watanabe, K. *Curr. Opin. Biotechnol.* **2001**, *12*, 237-241.
2. Watanabe, K.; Baker, P. W. *J. Biosci. Bioeng.* **2000**, *89*, 1-1 1.
3. Head, I. M.; Bailey, M. J. *Curr. Opin. Biotechnol.* **2003**, *14*, 245-247.
4. Davis, K. E. R.; Joseph, S. J.; Janssen, P. H. *Appl. Environ. Microbiol.* **2005**, *71*, 826-834.
5. Brockman, F. J. *Mol. Ecol.* **1995**, *4*, 567-578.
6. Stapleton, R. D.; Ripp, S.; Jimenez, L.; Cheol-Koh, S.; Fleming, J. T.; Gregory, I. R.; Sayler, G. S. *J. Microbiol. Methods.* **1998**, *32*, 165-178.
7. Fortin, N.; Beaumier, D.; Lee, K.; Greer, C. W. *J. Microbiol. Methods.* **2004**, *56*, 181-191.
8. Zhou, J.; Bruns, A. M.; Tiedje, J. M. *Appl. Environ. Microbiol.* **1996**, *62*, 316-322.
9. Ogram, A.; Sayler, G. S.; Barkay, T. *J. Microbiol. Methods.* **1987**, *7*, 57-66.
10. Picard, C.; Ponsonnet, C.; Paget, E.; Nesme, X.; Simonet, P. *Appl. Environ. Microbiol.* **1992**, *58*, 2717-2722.
11. Plaza, G.; Ulfig, K.; Hazen, T. C.; Brigmon, R. L. *Acta Microbiol. Pol.* **2001**, *50*(3-4), 205-218.
12. Bertrand, H.; Poly, R.; Van, V. T.; Lombard, N.; Nalin, R.; Vogel, T. M.; Simonet, P. *J. Microbiol. Methods.* **2005**, *62*, 1-11.
13. Mumy, K. L.; Findlay, R. H. *J. Microbiol. Methods.* **2004**, *57*, 259-268.
14. Glick, B. R.; Pasternak, J. J. *Molecular biotechnology-principles and applications of recombinant DNA;* 2nd ed.; ASM Press: Washington, DC, 1998.
15. Saleh-Lakha, S.; Miller, M.; Campbell, R. G.; Schneider, K.; Elahimanesh, P.; Hart, M. M.; Trevors, J. T. *J. Microbiol. Methods.* **2005**, *63*, 1-19.
16. Beller, H. R.; Kane, S. R.; Legler, T. C.; Alvarez, P. J. *J. Environ. Sci. Technol.* **2002**, *36*, 3977-3984.
17. Ritalahti, K. M.; Löffler, F. E. *Appl. Environ. Microbiol.* **2004**, *70*(7), 4088-4095.
18. Wellington, E. M. H.; Berry, A.; Krsek, M. *Curr. Opin. Microbiol.* **2003**, *6*, 295-301.
19. Holmes, D. E.; Nevin, K. P.; Lovley, D. R. *Appl. Environ. Microbiol.* **2004**, *70*(12), 7251-7259.
20. Schramm, A.; Amann, R. Nucleic acid-based techniques for analyzing the diversity, structure, and dynamics of microbial communities in wastewater

treatment; *Environmental Processes I . Wastewater Treatment,* 2nd ed.; Winter, J., editor; Wiley-VCH: Weinheim, 2001; Chapter 5, pp. 85-108.

21. Kirk, J. L.; Beaudette, L. A.; Hart, M.; Moutoglis, P.; Klironomos, J. N.; Lee, H.; Trevors, J. T. *J. Microbiol. Methods.* **2004,** *58,* 169-188.

22. Dojka, M. A.; Hugenholtz, P.; Hank, S. K.; Pace, N. R. *Appl. Environ. Microbiol.* **1998,** *64*(10), 3869-3877.

23. Hou, L.-H.; Dutta, S. K. *Lett. Appl. Microbiol.* **2000,** *30,* 238-243.

24. Peel, M.; Wyndham, R. C. *Appl. Environ. Microbiol.* **1999,** *65*(4), 1627-1635.

25. Marsh, T. L. *Curr. Opin. Microbiol.* **1999,** *2,* 323-327.

26. Marsh, T. L.; Saxman, P.; Cole, J.; Tiedje, J. M. *Appl. Environ. Microbiol.* **2000,** *66*(8), 36, 16-3620.

27. Cole, J. R.; Chai, B.; Farris, R. J.; Wang, Q.; Kulam, S. A.; McGarrell, D. M.; Garrity, G. M.; Tiedje, I. M. *Nucelic Acids Res.* **2005,** *33,* 294-296.

28. Petrie, L.; North, N. N.; Dollhopf, S. L.; Balkwill, D. L.; Kostka, J. E. *Appl. Environ. Microbiol.* **2003,** *69*(12), 7467-7479.

29. Macbeth, T. W.; Cummings, D. E.; Spring, S.; Petzke, L. M.; Sorenson, K. S., Jr. *Appl. Environ. Microbiol.* **2004,** *70*(12), 7329-7341.

30. Katsivela, E.; Moore, E. R. B.; Maroukli, D.; Strömpl, C.; Pieper, D.; Kalogerakis, N. *Biodegradation.* **2005,** *16,* 169-180.

31. Muyzer, G.; de Waal, E. C.; Uitterlinden, A. G. *Appl. Environ. Microbiol.* **1993,** *59*(3), 695-700.

32. Iwamoto, T.; Nasu, M. *J. Biosci. Bioeng.* **2001,** *92*(1), 1-8.

33. Baker, P.; Futamata, H.; Harayama, S.; Watanabe, K. *Environ. Microbiol.* **2001,** *3*(3), 187-193.

34. Singh, B. K.; Walker, A.; Morgan, J. A.; Wright, D. J. *Appl. Environ. Microbiol.* **2003,** *69*(9), 5198-5206.

35. Yang, Y.; Pesaro, M.; Sigler, W.; Zeyer, J. *Water Research* **2005,** *39,* 3954-3966.

36. Röling, W. F. M.; Milner, M. G.; Jones, D. M.; Fratepietro, F.; Swannell, R. P. J.; Daniel, F.; Head, 1. M. *Appl. Environ. Microbiol.* **2004,** *70*(5), 2603-2613.

37. Kleikemper, J.; Schroth, M. H.; Sigler, W. V.; Schmucki, M.; Bernasconi, S. M.; Zeyer, J. *Appl. Environ. Microbiol.* **2002,** *68*(4), 1516-1523.

38. Feris, K. P.; Hristova, K.; Gebreyesus, B.; Mackay, D.; Scow, K. M. *Microbiol. Ecol.* **2004,** *48*(4), 589-600.

39. Fjellbirkeland, A.; Torsvik, V.; Ovreas, L. *Antonie Van Leeuivenhoek.* **2001,** *79*(2), 209-217.

40. Knief, C.; Vanitchung, S.; Harvey, N. W.; Conrad, R.; Dunfield, P. F.; Chidthaisong, A. *Appl. Environ. Microbiol.* **2005,** *71*(7), 3826-3831.

148

41. Henckel, T.; Friedrich, M.; Conrad, R. *Appl. Environ. Microbiol.* **1999**, *65*, 1980-1990.
42. Gray, N. D.; Head, I. M. *Environ. Microbiol.* **2001**, *3*(8), 481-492.
43. Amann, R.; Fuchs, B. M.; Behrens, S. *Curr. Opin. Biotechnol.* **2001**, *12*, 231-236.
44. Lipski, A.; Friedrich, U.; Altendorf, K. *Appl. Microbiol. Biotechnol.* **2001**, *56*(1-2), 40-57.
45. Watanabe, K.; Watanabe, K.; Kodama, Y.; Syutsubo, K.; Harayama, S. *Appl. Environ. Microbiol.* **2000**, *66*(11), 4803-4809.
46. Christensen, N.; Batstone, D. J.; He, Z.; Angelidaki, I.; Schmidt, J. E. *Water Sci. Technol.* **2004**, *50*(9), 237-244.
47. Yang, Y.; Zeyer, J. *Appl. Environ. Microbiol.* **2003**, *69*(5), 2879-2883.
48. Aulenta, F.; Rossetti, S.; Majone, M.; Tandoi, V. *Appl. Microbiol. Biotechnol.* **2003**, *64*(2), 206-212.
50. Wackett, L. P. *Trends Biotechnol.* **2004**, *22*(4), 153-154.
51. Radajewski, S.; Ineson, P.; Parekh, N. R.; Murrell, J. C. *Nature.* **2000**, *403*, 646-649.
52. Radajewski, S.; McDonald, I. R.; Murrell, J. C. *Curr. Opin. Biotechnol.* **2003**, *14*, 296-302.
53. Radajewski, S.; Webster, G.; Reay, D. S.; Morris, S. A.; Ineson, P.; Nedwell, D. B.; Prosser, J. I.; Murrell, J. C. *Microbiol.* **2002**, *148*, 2331-2342.
54. Dumont, M. G.; Murrell, J. C. *Nat. Rev. Microbiol.* **2005**, *3* (6), 499-504.
55. Singleton, D. R.; Powell, S. N.; Sangaiah, R.; Gold, A.; Ball, L. M.; Aitken, M. D. *Appl. Environ. Microbiol.* **2005**, *71*(3), 1202-1209.
56. Jeon, C. O.; Park, W.; Padmanabhan, P.; DeRito, C.; Snape, J. R.; Madsen, E. L. *Proc. Natl. Acad. Sci. USA.* **2003**, *100*, 13591-13596.
57. Manefield, M.; Whiteley, A. S.; Griffiths, R. I.; Bailey, M. J. *Appl. Environ. Microbiol.* **2002**, *68*, 5367-5373.
58. Mahmood, S.; Paton, G. I.; Prosser, J. I. *Environ. Microbiol.* **2005**.
59. Cook, K. L.; Sayler, G. S. *Curr. Opin. Biotechnol.* **2003**, *14*, 311-318.
60. Rhee, S-K.; Liu, X.; Wu, L.; Chong, S. C.; Wan, X.; Zhou, J. *Appl. Environ. Microbiol.* **2004**, *70*(7), 4303-4317.
61. Schloss, P. D.; Handelsman, J. *Curr. Opin. Biotechnol.* **2003**, *14*, 303-310.
62. Beja, O.; Aravind, L.; Koonin, E. V.; Suzuki, M. T.; Hadd, A.; Nguyen, L. P.; Jovanovich, S. B.; Gates, C. M.; Feldman, R. A.; Spudich, J. L.; et al. *Science.* **2000**, *289*, 1902-1906.
63. Beja, O.; Suzuki, M. T.; Heidelberg, J. F.; Nelson, W. C.; Preston, C. M.; Hamada, T.; Eisen, J. A.; Fraser, C. M.; Delong, E. F. *Nature.* **2002**, *415*, 630-633.

64. Schloss, P. D.; Handelsman, J. *Curr. Opin. Biotechnol.* **2003**, *14,* 303-310.
65. Liu, W.-T.; Zhu, L. *Trends Biotechnol.* **2005**, *23*(4), 174-179.
66. de Hoog, C. L.; Mann, M. *Annu. Rev. Genomics Hum. Genet.* **2004**, *5*, 267-293.
67. Graves, P. R.; Haystead, T. A. J. *Microbiol. Mol. Biol. Rev.* **2002**, *66*(1), 39-63.
68. Werf, M. J.; Jellema, R. H.; Hankemeier, T. *J. Ind. Microbiol. Biotechnol.* **2005**, *32*(6), 234-252.
69. Mayeno, A. N.; Yang, R. S. H.; Reisfeld, B. *Environ. Sci. Technol.* **2005**, *39*(14), 5363-5371.

Modeling in Physical, Chemical, and Biological Methods

Chapter 9

Modeling Biodegradation and Reactive Transport: Analytical and Numerical Models

Yunwei Sun and Lee Glascoe[*]

Lawrence Livermore National Laboratory, 7000 East Avenue, Livermore, CA 94550

The computational modeling of the biodegradation in contaminated groundwater systems accounting for biochemical reactions coupled to contaminant transport is a valuable tool for both the field engineer/planner with limited computational resources and the expert computational researcher less constrained by time and computer power. There exists several analytical and numerical computer models that have been and are being developed to cover the practical needs put forth by users to fulfill this spectrum of computational demands. Generally, analytical models provide rapid and convenient screening tools running on very limited computational power, while numerical models can provide more detailed information with consequent requirements of greater computational time and effort. While these analytical and numerical computer models can provide accurate and adequate information to produce defensible remediation strategies, decisions based on inadequate modeling output or on over-analysis can have costly and risky consequences. In this chapter we consider both analytical and numerical modeling approaches to biodegradation and reactive transport. Both approaches are discussed and analyzed in terms of achieving bioremediation goals, recognizing that there is always a tradeoff between computational cost and the resolution of simulated systems.

Introduction

It has long been recognized that quantitative tools are necessary for the assessment and management of natural attenuation and bioremediation (for example, see (1)). Typically, mathematical models are used to determine the development of subsurface contaminant plumes and to evaluate the effectiveness of different bioremediation strategies. A wide range of numerical and analytical computer codes is currently available (2), which can be used to solve those biodegradation and reactive transport mathematical models. To appropriately select and apply computer codes, it is necessary to understand (a) the modeled physical system, (b) the modeling scope, (c) the computer code assumptions and limitations, and (d) the tradeoff between modeling resolution and computational cost. Reactive bioattenuation examples include processes such as radionuclide decay, denitrification, biodegradation of chlorinated solvents, etc. Consider the anaerobic degradation of the most frequently detected organic compounds in groundwater, perchloroethylene (PCE) and trichloroethylene (TCE) (3); the reaction pathway of PCE and TCE biodegradation is shown in Figure 1 (4, 5). PCE reacts to produce TCE; TCE reacts to produce three daughter species, cis-1,2-dichloroethylene (cis-1,2-DCE), trans-1,2-dichloroethylene (trans-1,2-DCE), and 1,1-dichloroethylene (1,1-DCE), simultaneously; the three daughter species further react to produce vinyl chloride (VC); and VC reacts to produce ethylene (ETH) (3, 4).

In order to simulate the transport phenomena with the coupled PCE/TCE reaction network, the general mass balance equations need to be solved (6):

$$\frac{\partial c_i}{\partial t} = -\nabla \cdot \left(\mathbf{v}_i c_i - \nabla \cdot \mathbf{D}_i c_i \right) + \frac{q_i^s c_i^s}{\phi} + f_i(\mathbf{c}), \quad \forall i = 1, 2, \cdots, n \qquad (1)$$

where c_i [M L^{-3}] is the concentration of ith species; t is time [T]; $\mathbf{v}$ [LT^{-1}] is the vector of velocity; $\mathbf{D}$ [L^{2}T^{-1}] is the tensor of dispersion coefficients; n is the total number of species; q^s and c^s are the source/sink quantity and concentration; and f is the gain or loss due to reactions. The source/sink term can also be written as boundary conditions. Eq 1 can be solved either analytically or numerically based on the complexity of the modeled systems. The discussion below focuses on how the modeling of complex reactive systems, such as PCE and TCE degradation, would be approached for both analytical and numerical solutions.

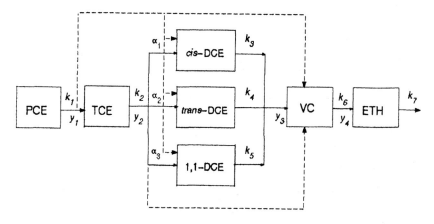

Figure 1. PCE and TCE degradation pathway. k_i, $\forall i = 1,2,\cdots, 7$ are the first-order reaction rates; y_i, $\forall =1,2,3,4$ are yield coefficients; and α_1, α_2, α_3, are product distribution factors of the reduction of trichloroethylene (TCE). Dashed lines represent other possible reaction

Analytical Modeling of Biodegradation and Reactive Transport

Scope of Analytical Models

After constructing an appropriate mathematical model in terms of relevant state variables for problems of interest, the model must be solved either directly through analytical means or by employing numerical methods. For the sake of speed and efficiency, the preferred method to solve the model is an analytical solution; however, most problems of practical interest introduce complexities, such as irregular shapes of domain-boundaries, heterogeneities, non-linearities, and irregular source functions, which constrain the derivation of an analytic solution. For this reason, numerical methods are employed to solve the mathematical model, thus increasing computational effort that increases with modeling resolution.

The use of numerical and analytical solutions should be viewed as mutually complementary. Sometimes, the complexity of contaminant systems may require the use of numerical models to represent, for example, a special geometry, heterogeneity, or distributed physical and chemical properties. However, computational complexity resulting from the implementation of physically and geometrically intricate aspects of a simulation model results in increased algorithm detail and a more involved algorithm execution. In addition to increased simulation demands, higher labor costs for model development and maintenance and data collection requirements are important concerns, especially for numerical models. Because of the advantages of analytical solutions (including efficiency, speed and ease-of-use), it is often preferable to simplify a given flow and transport problem to the extent that an analytical solution can be obtained.

Often rapid analytical solutions are used to facilitate confident decision-making, and numerical solutions are applied at more advanced stages of bioremediation problem at higher resolution levels. Also, analytical models are required for the verification of the development of numerical codes, which may solve very complex problems. When deciding between the tradeoffs of model resolution and computational effort and, therefore, between the employment of an analytical model as opposed to a numerical model, one should consider the five reasons to use analytical solutions outlined by Javandel et al. (7):

- Where applicable, analytical methods are the most economical approach.
- Analytical methods are probably the most efficient alternative when data necessary for identification of the system are sparse and uncertain.
- Analytical methods are always the most useful means for an initial estimation of the order of magnitude of contaminant extent.
- Analytical models do not require experienced modelers or complex numerical codes.
- Analytical models, in many cases, provide a rough estimate that can be obtained through published tables. When application of simple computer codes for evaluation of analytical solutions is needed, the input data are usually very simple and do not require a detailed familiarity with the codes.

Discussed in the subsections below are specific examples of how reactive transport can be represented and solved analytically. Note that while we are not limiting ourselves in this discussion to a one-dimensional description, in order to simplify the presentation of the mathematical model, only a one-dimensional case is discussed.

Analytical Models and Solutions

Analytical solutions of reactive transport are usually developed under relatively simple flow and transport conditions. The differential equations (eq 1) governing species transport with first-order reactions in a groundwater system are described as (8):

$$\frac{\partial \mathbf{c}}{\partial t} + L(\mathbf{c}) = \mathbf{Ac}, \quad L(\mathbf{c}) = -D\frac{\partial^2}{\partial x^2} + v\frac{\partial}{\partial x} \tag{2}$$

where $\mathbf{c}$ is the vector of concentrations $[ML^{-3}]$; D is the dispersion coefficient $[L^2T^{-1}]$; v is the groundwater velocity $[LT^{-1}]$; x is the coordinate in the direction of flow $[L]$; and $\mathbf{A}$ is the first-order reaction matrix.

If vector $\mathbf{c}$ becomes a scalar variable c, eq 2 represents the transport system of a single species. The reaction matrix is determined by the reaction networks (or biodegradation pathways), which can be sequential, parallel, reversible, or convergent as illustrated in Figure 2.

To best review the capabilities along with the necessary assumptions and limitations of modeling reactive transport using analytical solutions, we discuss three topical modeling examples: single species transport, multiple species transport, and reactive transport in fractured rock.

Single Species Transport

Ogata (9) and Bear (10) were the first to derive analytical solutions to contaminant transport equations for one-dimensional problems. Ogata's solution covers advection and dispersion, while Bear's solution added the first-order reaction for a single contaminant species. Since these pioneering efforts, the development of analytical solutions for contaminant transport problems has become an important part of contaminant hydrology, van Genuchten and Alves (11) and Toride et al. (12) compiled various analytical solutions for solving the one-dimensional solute transport equations.

Beljin (13) and Wexler (14) reviewed analytical solute transport models for one-, two-, and three-dimensional systems. However, for all of these solutions, the fundamental partial differential equations (PDEs) represent the transport of either a nonreactive tracer or a single reactive species. Reactions are assumed to be first-order in these solutions and are used primarily to address transport phenomena, rather than biochemical reactions. When these solutions are used to represent contaminant transport, the reaction products from biodegradation cannot be addressed. Therefore, these analytical solutions are of the most limited use when addressing, for example, one of the most challenging aspects of managing chlorinated solvent contaminants: the fact that the degradation daughter products are often more toxic than the parent contaminant (*15*).

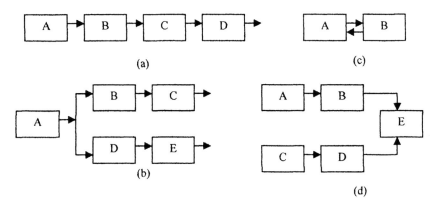

Figure 2. Basic reaction patterns: (a) sequential reactions; (b) parallel reactions; (c) reversible reactions; and (d) convergent reactions.

Multiple Species Transport

Since the 1970s there have been efforts to derive analytical solutions for the reactive transport of multiple species. Cho (*16*) derived an analytical solution to a three-species chain with a simple boundary condition. Lunn *et al* (*17*) used the Fourier transform method to derive the same solution of three-species chain with first-order reactions. van Genuchten (*18*) used the Laplace transform method and extended previous analytical solutions to a four-species chain. Because of the complexity of inverse Laplace transforms or other integral transforms, the difficulty of deriving analytical solutions for the transport of multiple sequentially reactive species increases exponentially with the species number. Sometimes, implementation of analytical solutions becomes a tedious process because complicated mathematics is involved, and, perhaps worse, there is a perceived gap between the development of analytical solutions and practical applications.

In order to bridge this perceived "practicality gap" for analytical solutions, Sun *et al.* (*19*) derived a linear transform approach, which decomposes a reactive transport system coupled by reaction terms into multiple independent subsystems with easily derived analytical solutions. The transform is defined as a linear function of species concentrations as (*19*):

$$a_i = c_i + \sum_{j=1}^{i-1} \left[\prod_{l=j}^{i-1} \frac{y_l k_l}{k_l - k_i} \right] c_j, \qquad \forall i = 1, 2, \cdots, n \qquad (3)$$

where a_i [ML^{-3}] is the concentration of species i in the transformed domain ("a" domain). Then, the system equation coupled by reaction matrix **A** can be simplified as:

$$\frac{\partial \mathbf{a}}{\partial t} + L(a) = \Lambda \mathbf{a} \qquad (4)$$

where Λ is a diagonal matrix of $n \times n$ containing the first-order reaction rates of all species, k_i, $i=1,2,\cdots,n$. By employing this transform, each partial differential equation (PDE) in eq 4 becomes independent and may be solved separately. The transform significantly reduces the complexity of the coupled reactive transport system (eq 2), and, hence, analytical solutions can either exist or can be developed. The solution scheme of Sun et al. (19) extended the analytical solution capabilities from four species to an unlimited number of species. However, this method is only limited to serial reaction patterns or the parallel reaction networks that can be further decomposed to serial patterns. Sun et al. (20) extended this work by deriving an analytical solution of first-order decay chain in multiple phases using this transformation.

Clement (21) mathematically proved Sun et al.'s (19) transform and illustrated that singular value decomposition (SVD) can be used to derive analytical solutions of transport with convergent reactions. When SVD is used, analytical solutions become available for convergent and reversible reactions. For instance, if the reaction matrix, **A**, is a diagonalizable matrix, it can be written as:

$$\mathbf{A} = \mathbf{S}\Lambda\mathbf{S}^{-1} \qquad (5)$$

where Λ is a diagonal matrix containing the eigenvalues of **A**, and **S** is a matrix whose columns are linearly independent eigenvectors of **A**. For the sequential first-order reactions, Λ in eq 5 is exactly the diagonal matrix Λ in eq 4. Substituting eq 5 into eq 2 yields

$$\frac{\partial \mathbf{c}}{\partial t} + L(\mathbf{c}) = \mathbf{S}\Lambda\mathbf{S}^{-1}\mathbf{c} \qquad (6)$$

Multiplying by $\mathbf{S}^{-1}$, eq 6 becomes eq 4 where $\mathbf{a}=\mathbf{S}^{-1}\mathbf{c}$. Each PDE in eq 4 is independent of other PDEs. The analytical solutions of eq 4, in terms of **a**, are available for a variety of boundary conditions. Finally, the solution of **c** can be derived as **C = S a.**

Clement (*21*) solved the SVD method numerically for the transform matrices while the solutions were derived semi-analytically. Lu *et al.* (*8*) derived a closed-form solution of TCE transport with the convergent reactions using SVD. The analytical solution of Lu *et al.* (*8*) demonstrated the significance of considering the convergent reactions. Sun *et al.* (*20*) extended Lu *et al.'s* work and derived analytical solutions for the entire PCE reaction network (Figure 1).

Note that all of the linear systems discussed above (*19, 21, 8, 20*) have a fundamental assumption that all represented species share the same retardation factor. Eykholt and Li (*22*) developed a semi-analytical solution of a linear reaction network with different retardation factors using a response function approach. Since numerical convolution is involved, the method is less transparent, more difficult to employ, and, ultimately, difficult to implement as a screening/planning tool. Bauer *et al.* (*23*) developed a Laplace domain solution using a recursive form for a different retardation system. The concentration of a daughter species is expressed as a linear function of ancestor concentrations, and the factor of each species concentration is calculated using the recursive form. Quezada *et al.* (*24*) used Laplace transformation and linear transformation to decouple the first-order reactive system with distinct retardation factors. Though both Bauer *et al.* (*23*) and Quezada *et al.* (*24*) have made significant progress in first-order reactive transport, the complexity of inverse Laplace transforms makes code implementation difficult. When numerical inverse Laplace transforms are involved, the approach becomes even more complicated. Both the Eykholt and Li (*22*) and the Bauer *et al.* (*23*) approaches are based on a unimolar assumption, that is, the stoichiometry of the reaction is such that one mole of product is produced by consuming one mole of reactant.

Reactive Transport in Fractured Rock

The potential geologic repository for high-level nuclear waste at Yucca Mountain, much of which is comprised of a fractured volcanic tuff, has led to an increased interest in the behavior of radionuclide transport in fractures and in better understanding ground water flow and transport in fractured rock (e.g., *25, 26*). Starting from a single or from a few parallel fractures in the rock matrix, several studies have focused on deriving analytical solutions. Tang *et al.* (*27*) developed an analytical solution to a mathematical model and investigated the fundamental dynamics of the transport of a single radionuclide along a single fracture. Sudicky and Frind (*28*) extended this solution to multiple parallel fractures, and, later, they derived an analytical solution for the reactive transport of a two-member decay chain in the single fracture (29). Cormenzana (30) provided a simplified form of the Sudicky and Frind (29) solution. Because of the difficulties involved in inverse Laplace transforms, analytical solutions for transport in fractured systems are limited to one or two species. To avoid the difficulties noted by Sudicky and Frind (29) in extending their approach to multiple species, the transformation of Sun *et al.* (19) was employed to decompose the partial differential equations, which are coupled by reaction terms into independent subsystems. Then, the solution to single species transport in the single fractured system can be applied to derive solutions to N-member solvable decay chains (31).

Numerical Modeling of Biodegradation and Reactive Transport

Analytical and numerical models of biodegradation and reactive transport are always complementary. Compared with analytical solutions, numerical methods provide inexact solutions with fewer simplifying assumptions. Additionally, numerical solutions can be used to simulate transport within complex hydrogeologic systems coupled with multiple reactions for which reaction stoichiometry and networks vary spatially and temporally. The hydrogeologic systems can be heterogeneous and anisotropic, and flow systems can be transient with complex initial and boundary conditions. Numerical methods have been successfully and routinely employed for simulating biodegradation and reactive transport (32, 33, 34, 35, 36, 37, 2, 38, 39, 40).

Operator-split Numerical Approach

Operator-split strategy is the best practical approach (41) to solve the partial differential equations coupled by reactions, although many other numerical solution schemes exist. In the operator-split approach, the partial-differential-equation system (eq 1) can be split into a few subsystems (Table I): advection, dispersion, source/sink-mixing, and reactions, which are then solved using appropriate approaches (35, 41).

In the basic reactive transport equation (eq 1), advection, dispersion, and source/sink-mixing are species-independent. The velocity and dispersion coefficient are normalized by the retardation factor of the species. The first three terms in Table I for each species can be solved without considering other species concentrations. However, the reaction equations are location-independent and

Table I. Operator-split Components

PDEs	Advection	Dispersion	Source-Sink Mixing	Reactions
$\dfrac{\partial c_i}{\partial t} =$	$-\nabla \cdot \mathbf{v}_i c_i$	$\nabla \cdot \left(\nabla \cdot \mathbf{D}_i c_i \right)$	$\dfrac{q_i^s c_i^s}{\phi}$	$f(c)$

$$\forall i = 1, 2, \cdots, n$$

are coupled by other species concentrations at the same location. In other words, reaction equations can be solved only by discretizing the time domain.

Much of the difficulty of modeling reactive transport and biodegradation is associated with the representation of the often complicated reaction term, f(c). Discussed below are some of the specific issues associated with solving common numerical problems in reactive transport modeling including the modeling of reaction kinetics.

Numerical Modeling of Reaction Kinetics

The partial differential equations described by eq 1 are coupled by the reaction term, $f(c)$. Depending on the type of system being analyzed, various mathematical forms of reaction kinetics can be used to describe this term (*35, 33*). For example, depending upon the prevailing environmental conditions, aerobic biodegradation can be described as (a) no reaction, (b) a first-order reaction, (c) an instantaneous reaction, (d) a Monod reaction with constant biomass, or (e) a single or dual Monod reaction coupled to biomass growth.

No Reaction

Often, it is assumed that subsurface contaminants are stable, i.e., they do not interact with subsurface systems or degrade in any way. In such a case, the reaction term, $f(c)$, is set to zero, and a single-species transport system is used to provide a basis for demonstrating bioremediation.

First-order Reaction

The first-order reaction is expressed as the increased amount of a contaminant by the decay (degradation) of its parent species and the decreased amount by its decay into the daughter species:

$$\frac{\partial c_i}{\partial t} = k_{i-1}c_{i-1} - k_i c_i \qquad (7)$$

with the k_i canceling for $i=1$ and $i=n$. The analytical solution to the system of eq 7 for batch reactor conditions has been in use for nearly 100 years (*42*). However, numerical solvers are required when transport is involved, especially when eq 7 becomes "stiff" (described in further detail below).

Instantaneous Reaction

In many instances the parameters that describe the kinetics of biological reactions are not available, and the aerobic biodegradation process is approximated by an instantaneous reaction (*43*). The ratio of the amount of oxygen consumed to the amount of contaminant destroyed by the reaction is usually estimated by an appropriate stoichiometric constant, and the change of concentration due to the instantaneous reaction is expressed as:

$$\Delta c = o / F, o = 0, \forall c > o / F \qquad \Delta o = cF, c = 0, \forall c < o / F$$

where c and o are the contaminant and oxygen concentrations and F is the ratio of oxygen to contaminant consumed (*34*).

Reaction with Constant Biomass Concentration

The double Michaelis-Menten law (*43, 44, 34, 45, 46, 2, 47*) is often used to describe the kinetics of microbial transformation and nutrient (oxygen) and substrate (contaminant) concentrations. If a dual-substrate Monod expression is used, assuming the biomass concentration is constant everywhere at any time, the nonlinear reaction term can be written for the contaminant and oxygen concentrations as:

$$f_c = X\mu\left(\frac{c}{K_c + c}\right)\left(\frac{o}{K_o + o}\right), f_o = F f_c$$

where μ is the maximum contaminant utilization rate per unit biomass $[T^{-1}]$; X is the constant biomass concentration $[M\ L^{-3}]$; and K_c and K_o are contaminant and oxygen half saturation constants $[M\ L^{-3}]$, respectively. Sometimes, a multiple Monod equation is used (48, 37, 47, 49):

$$f_c = X\mu\prod_{i=1}^{n}\left(\frac{c_i}{K_i + c_i}\right)$$

Reaction With Biomass Growth

To fully describe the biodegradation processes, microbial growth and transport must also be considered. The full coupling of Monod kinetics with biomass growth, decay, and attachment/detachment between the liquid and solid phases can be expressed as (33):

$$f_c = \mu\left(X_a + \frac{\rho X_s}{\phi}\right)\left(\frac{c}{K_c + c}\right)\left(\frac{o}{K_o + o}\right),\ f_o = Ff_c$$

$$fx_a = \mu y X_a\left(\frac{c}{K_c + c}\right)\left(\frac{o}{K_o + o}\right) - K_r X_a - K_a X_a + \frac{K_d \rho X_s}{\phi}$$

$$f_x = \mu y X_s\left(\frac{c}{K_c + c}\right)\left(\frac{o}{K_o + o}\right) - K_r X_s - K_d X_s + \frac{K_a \phi X_a}{\rho}$$

where X_a, $[M\ L^{-3}]$ and X_s $[M\ M^{-1}]$ are the aqueous phase and solid phase biomass concentrations, respectively; y is biomass/substrate yield coefficient $[M\ M^{-1}]$; K_r is decay rate $[T^{-1}]$; K_a and K_d are attachment and detachment coefficients $[T^{-1}]$; and ρ is bulk density $[M\ L^{-3}]$. Note that biomass is exchanged between aqueous phase and solid phase, and the solid phase biomass is often assumed to not be transported by either advection or dispersion.

Coupled Stiff Reactions

Unlike equilibrium reactions, which can be expressed in a generalized format of stoichiometric matrices, kinetic reactions vary from linear to nonlinear formats and from nonstiff to stiff problems. Often, the formulation of reaction kinetics is site-specific, even varying with time and spatial coordinates. When kinetics reactions are coupled in the partial differential equations for transport, the operator-split strategy is often utilized to develop a general solution scheme.

Incorporation of detailed kinetic reactions in a transport system can result in a "stiffness" of governing equations. As a consequence of this equation stiffness, explicit time integration of the reaction source terms is restricted to very small time steps. The system is said to be "stiff" when two species have very different natural time scales. Environmental and geochemical systems are often stiff when equilibrium (fast reactions) and kinetics (slow reactions) are coupled in transport equations. However, fully implicit time integration of reaction source/sink terms together with implicit time schemes of transport equations requires much larger computer memories. It is unnecessary, or at least inefficient, to use stiff ODE (ordinary differential equation) solvers, such as implicit time integration of reaction source/sink terms, to solve non-stiff systems. In order to cover a wide range of possible reactions (fast or slow, stiff or nonstiff) for groundwater biodegradation, it is essential to measure stiffness and select the appropriate solvers. An efficient operator split construction is used in RT3D (*35*) to solve the coupled transport equation employing the LSODE/LSODI solver (*50*) for solving the reaction terms.

Kinetics reaction terms in the transport PDEs cannot be explicitly treated as source/sink terms. In order to satisfy the stability condition of ODE solutions, Δt for the reaction terms must be very small. In other words, each transport time step needs to be subdivided into many reaction time steps. However, when the ODEs of the reaction terms become stiff, explicit solution schemes will fail. One can get the practical feeling about the stiffness from the following first-order reactions when $k_1 = 10^6$ and $k_2 = 10^{-3}$:

$$C_1 \xrightarrow{k_1} C_2 \xrightarrow{k_2}$$

The reaction terms are expressed as:

$$\frac{\partial c_1}{\partial t} = -k_1 c_1, \quad \frac{\partial c_2}{\partial t} = -k_1 c_1 - k_2 c_2$$

The stability condition requires $\Delta t < 10^{-6}$ while the influence of $k_2 c_2$ can only be seen after $10^3 \sim 10^9$ time steps. In this way, the explicit scheme will cause the accumulation of round-off error. Gear (*51*) developed the first algorithm for such stiff ODEs. One can use MATLAB (*52*) ODE solvers, ode45 and ode45s for this problem.

When both equilibria and kinetics are coupled in the transport equations, flexible and robust solvers need to be selected for handling the mixed systems of differential and algebraic equations. Clement *et al.* (*35*) successfully implemented LSODE (*50*) into RT3D to cover a wide range of kinetic reactions. However, the current version of RT3D is only limited to kinetic reactions. To include both equilibria and kinetics, it is suggested that both LSODE and LSODI be implemented.

Representative Analytical and Numerical Codes

Analytical Computer Codes for Biodegradation and Reactive Transport

Several computer codes have been developed based on analytical solutions of reactive transport. This limited review only considers those frequently used in the environmental remediation industry.

AT123D

AT123D (53) is a three-dimensional analytical solution package for transport and fate. Processes simulated include advection, dispersion, diffusion, adsorption, and biological decay. Contaminant releases can be simulated as instantaneous, continuous or varying loads. Source load configurations can be established as a point, line, plane, or volume release.

BIOSCREEN

BIOSCREEN is a screening level transport code that simulates the bioremediation and reactive transport of dissolved hydrocarbons or many other contaminants. It is based on the Domenico (54) analytical solution of a single contaminant transport in three dimensions and can simulate advection, dispersion, adsorption (linear equilibrium) and first-order reaction. Groundwater flow is assumed uniform with constant velocity. The reaction options coupled in the contaminant transport include (1) no reaction, (2) first-order decay, and (3) biodegradation based on instantaneous reactions.

BIOCHLOR

BIOCHLOR (55) is a screening code that simulates remediation by natural attenuation of dissolved solvents at chlorinated solvent release sites. BIOCHLOR can be used to simulate solute transport without decay and solute transport with biodegradation modeled as a sequential first-order process within one or two different reaction zones. The software, programmed in the Microsoft Excel (Microsoft Corporation, Redmond, WA) spreadsheet environment and based on the Sun et al. (19) transform and the Domenico (54) analytical solution of solute transport model, has the ability to simulate one-dimensional advection, three-dimensional dispersion, linear adsorption, and biotransformation via reductive dechlorination. Reductive dechlorination is assumed to follow a sequential first-order decay process. However, BIOCHLOR is not limited to the biodegradation of chlorinated solvents. It can also be used for simulating radionuclide transport, denitrification, etc. Case studies can be found in (*56*).

PLUME2D

PLUME2D (*57*) is an analytical code based on closed-form solutions of the non-conservative solute transport equation for instantaneous and continuous releases of a tracer in one or more source locations as presented by Wilson and Miller (*58*). The program uses superposition of solutions for individual sources to calculate the resulting concentration distribution for a tracer in a homogeneous, confined aquifer with uniform regional flow. The code evaluates the effects of solute advection and dispersion in an aquifer with multiple (up to *25*) fully penetrating sources. It is limited to retardation and first-order decay for a single species.

Interactive Models for Groundwater Flow and Solute Transport

Valocci *et al.* (*59*) developed Java code for interactive modeling of flow and reactive transport in one, two, and three dimensions. This code covers solutions for single species transport. Using any Internet web browser, analytical modeling can be conducted remotely using this application.

Numerical Computer Codes for Biodegradation and Reactive Transport

Several numerical computer codes have been developed to model biodegradation and reactive transport. The limited review to follow considers only those frequently used in the environmental industry.

NUFT

The NUFT code (*37*) is a numerical suite for modeling multiphase, multicomponent reactive transport under non-isothermal conditions. It simulates advection, dispersion, reactions (both kinetics and equilibria), thermal conduction, and radiation, etc. in multiple overlapped porous and non-porous media. It has been successfully used in modeling of bioventing and steam injection for groundwater remediation (*47*) and of hydrothermal behavior at nuclear disposal sites. Finite element and finite difference solution options, together with internal structured, external nonstructured, and multigrid meshes, are available.

TOUGHREACT

The TOUGHREACT code (*40*) was developed based on TOUGH2 for multiphase and multicomponent reactive transport. A variety of equilibrium reactions are covered, such as aqueous complexation, as dissolution/exsolution, mineral dissolution/precipitation, and cation exchange. TOUGHREACT has been successfully used for modeling hydrothermal and geochemical systems at nuclear disposal sites and for CO_2 disposal in deep aquifers.

RT3D

The RT3D code (*35*) is a MODFLOW family code and was developed specifically for modeling bioremediation and reactive transport in a single phase. Since LSODE was implemented for solving a wide range of ODEs, it solves reactive transport with stiff kinetics.

BIOPLUME

The BIOPLUME codes (*60*) were developed for simulating transport of a single and multiple hydrocarbons in two dimensions with oxygen-limited and reactant-limited bioreactions. Many other numerical codes for modeling biodegradation and reactive transport are reviewed by Wiedemeier *et al.* (*2*) and Rifai and Rittaler (*1*).

Analytical and Numerical Solutions, an Example and Comparison

As a demonstration of the differences and similarities of analytical and numerical modeling of reactive transport and biodegradation, we illustrate with an example that is tractable analytically. Consider the one-dimensional reactive transport system of Sun and Clement (*15*). A reaction branch of the reaction network ((*15*), Figure 1) is used here as a sequential reactive transport problem as for given reaction rates, $\mathbf{k}=[k_1\,k_2\,...\,k_5]^T=[0.05\ 0.03\ 0.01\ 0.005\ 0.002]^T$:

$$
\begin{array}{ccccc}
k_1 & k_2 & k_3 & k_4 & k_5 \\
\end{array}
$$
$$
C_1 \rightarrow C_2 \rightarrow C_3 \rightarrow C_4 \rightarrow C_5 \rightarrow
$$

Both analytical (*19*) and numerical (*35*) solutions were computed for a column of 500 m discretized using 50 evenly spaced nodal points. A uniform groundwater transport velocity of 0.5 m/d and a dispersion coefficient of 5.0 m^2/d were assumed. Initial conditions for all species were assumed to be zero.

The boundary conditions assumed are similar to those used in deriving the basic analytical solution (*6*),

$$a_i = \frac{a_i^o}{2}\exp\left(\frac{vx}{2D}\right)\left[\exp(-B_ix)\mathbf{erfc}\gamma_i^- + \exp(B_ix)\right.$$

where

$$B_i = \left(\frac{v^2}{4D^2}+\frac{k_i}{D}\right)^{1/2}, \quad \mathbf{erfc}(\eta) = 1 - \mathbf{erf}(\eta) = \frac{2}{\sqrt{\pi}}\int_{\eta}^{\infty}\mathbf{e}$$

$$\gamma_i^{-1} = \frac{x-\left(v+4k_iD\right)^{1/2}t}{2\left(Dt\right)^{1/2}}, \quad \gamma_i^+ = \frac{x+\left(v+4k_iD\right)^{1/2}t}{2\left(Dt\right)^{1/2}}$$

Using eq 3 or eq 5, the transformation matrices are obtained:

$$\mathbf{S} = \begin{bmatrix} 1 & 0 & 0 & 0 & 0 \\[1em] \dfrac{k_1}{k_2-k_1} & 1 & 0 & 0 & 0 \\[1em] \prod\limits_{l=2}^{3}\dfrac{k_{l-1}}{k_l-k_1} & \dfrac{k_2}{k_3-k_2} & 1 & 0 & 0 \\[1em] \prod\limits_{l=2}^{4}\dfrac{k_{l-1}}{k_l-k_1} & \prod\limits_{l=3}^{4}\dfrac{k_{l-1}}{k_l-k_2} & \dfrac{k_3}{k_4-k_3} & 1 & 0 \\[1em] \prod\limits_{l=2}^{5}\dfrac{k_{l-1}}{k_l-k_1} & \prod\limits_{l=3}^{5}\dfrac{k_{l-1}}{k_l-k_2} & \prod\limits_{l=4}^{5}\dfrac{k_{l-1}}{k_l-k_3} & \dfrac{k_4}{k_5-k_4} & 1 \end{bmatrix}$$

$$\mathbf{S}^{-1} = \begin{bmatrix} 1 & 0 & 0 & 0 & 0 \\[1em] \dfrac{k_1}{k_1-k_2} & 1 & 0 & 0 & 0 \\[1em] \prod\limits_{l-1}^{2}\dfrac{k_l}{k_l-k_3} & \dfrac{k_2}{k_2-k_3} & 1 & 0 & 0 \\[1em] \prod\limits_{l-1}^{3}\dfrac{k_l}{k_l-k_4} & \prod\limits_{l-2}^{3}\dfrac{k_l}{k_l-k_4} & \dfrac{k_3}{k_3-k_4} & 1 & 0 \\[1em] \prod\limits_{l-1}^{4}\dfrac{k_l}{k_l-k_5} & \prod\limits_{l-2}^{4}\dfrac{k_l}{k_l-k_5} & \prod\limits_{l-3}^{4}\dfrac{k_l}{k_l-k_5} & \dfrac{k_4}{k_4-k_5} & 1 \end{bmatrix}$$

A constant concentration condition ($c_1^o = 1.0$, $c_i^o = 0$, $\forall i > 1$) is set at the inlet and a free boundary condition is assumed at the outlet boundary. Note that all yield coefficients $y_i = 1, \forall i$ are assumed in this example. Concentration profiles are compared in Figure 3 when $t = 730$ d. The CPU time ratio between analytical solution and numerical solution is usually very small (in this 5-species case the ratio was 0.116 between the analytic solution CPU and the RT3D solution CPU). This specific example illustrates both the faster run time and the mathematical transparency of the analytical solution in contrast to the numerical solution.

In addition, since analytical solutions can be expressed in a functional format, they can be easily used for system identification (as inverse problems, (*12, 61, 62*)), sensitivity analyses (*8*), and numerical code verification (*15, 63, 64*). However, numerical solutions are much closer to the real world systems and are never limited to the assumptions that are used to derive analytical solutions.

Conclusions and Future Considerations

In this chapter, analytical and numerical approaches to modeling reactive transport and bioremediation were reviewed, and the tradeoffs and benefits of

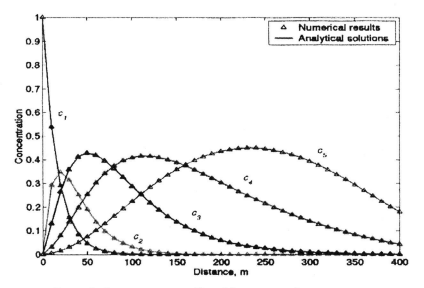

Figure 3. Concentration profiles of five species after two years.
(See page 2 of color inserts.)

both approaches were discussed. Because of their relative ease of use and speed in providing insight into biodegradation processes, analytical solutions play an important role in modeling bioremediation and natural attenuation at the screening and planning stages. However, when initial conditions, boundary conditions, and/or reaction kinetics require high-resolution, numerical models are often the best solution. For high-resolution modeling at late stages, analytical models can be used for numerical code verification and conceptual model validation. In all cases, n umerical a nd analytical m odels should b e v iewed as complementary tools, providing the planner and researcher with a dual-capability of making both rapid and detailed analysis of biodegradation and reactive transport systems.

A number of improvements to modeling of reactive transport are possible. It has been realized that the lack of the closed form analytical solution of sequentially reactive transport with different retardation factors has limited the application of analytical codes, such as BIOCHLOR (55). To sufficiently apply analytical solutions in real world systems, the conformal transformation is a promising direction for converting heterogeneous and anisotropic systems into uniform flow systems. Web-based simulation, especially for analytical solutions, has great potential. On the numerical side, it would be beneficial to develop a model translator, which can convert database information to model input files and convert model input files between different computer codes. Users may then benefit from the ability to select the most appropriate computer code and adjust (or easily update) the model input.

Acknowledgements

This work was performed under the auspices of the U. S. Department of Energy by the University of California, Lawrence Livermore National Laboratory under Contract No. W-7405-Eng-48.

References

1. Rifai, H.S.; Rittaler, T. *Biodegradation.* **2005**, *16*, 291–304.
2. Wiedemeier, T.H.; Rifai, H.S.; Newell, C.J.; Wilson, J.W. *Natural Attenuation of Fuels and Chlorinated Solvents;* John Wiley & Sons, Inc.: New York, 1999.
3. Wiedemeier, T.H.; Swanson, M.A.; Moutoux, D.E.; Gordon, E.K.; Wilson, J.T.; W ilson, B .H.; Kampbel, D.H.; Hansen, J.; Haas, P.; Chapelle, F.H. *Technical Protocol for Evaluating Natural Attenuation of Chlorinated Solvents in Groundwater;* Air Force Center for Environmental Excellence, Technology Transfer Division, Brooks AFB: San Antonio, TX, 1996.

4. Semprini, L.; Roberts, P. V.; Hopkins, G. D.; McCarty, P. L. *Ground Water.* **1990**, *28*, 715–727.

5. Sun, Y.; Lu, X.; Petersen, J.N.; Buscheck, T.A. *Transport Porous Med.* **2004**, *55*, 301–308.

6. Bear, J. *Groundwater Hydraulics;* McGraw-Hill: New York, 1979.

7. Javandel, I.; Doughty, C.; Tsang, C.F. *Groundwater Transport,* American Geographical Union: Washington, D.C., 1984.

8. Lu, X.; Sun, Y.; Petersen, J.N. *Transport Porous Med.* **2003**, *51*, 211–225.

9. Ogata, A. Ph.D. thesis, Northwestern University, Evanston, IL, 1958.

10. Bear, J. Ph.D. thesis, University of California, Berkeley, CA, 1960.

11. van Genuchten, M.Th.; Alves, W.J. *Analytical Solutions of the One-dimensional Convective-Dispersive Solute Equation;* Agricultural Research Service, Department of Agriculture: Riverside, CA, 1982.

12. Toride, N.; Leij, F.J.; van Genuchten, M.Th. *The CXTFIT Code for Estimating Transport Parameters from Laboratory and Field Tracer Experiments;* U.S. Salinity Laboratory Report: No. 137, 1995.

13. Beljin, M.S. *Review of three-dimensional Analytical Models for Solute Transport in Groundwater Systems;* International Ground Water Modeling Center, Holcomb Research Institute, Butler University: Indianapolis, IN, 1991.

14. Wexler, E.J., *Analytical Solutions for One-, Two-, and Three-dimensional Solute Transport in Ground-water Systems with Uniform Flow;* Book 3, Ch. B7, U.S. Geol. Survey: Denver, CO. 1992; p 199.

15. Sun, Y.; Clement, T.P. *Transport Porous Med.* **1999**, *37*, 327–346.

16. Cho, C.M. *Can. J. Soil Sci.* **1970**, *51*, 339–350.

17. Lunn, M.; Lunn, R.J.; Mackay, R. *J. Hydrology.* **1996**, *180*, 195–210.

18. van Genuchten, M.Th. *Comput. Geosci.* **1985**, *11*, 129–147.

19. Sun, Y.; Petersen, J.N.; Clement, T.P.; Skeen, R.S. *Water Resour. Res.* **1999**, *35*, 185–190.

20. Sun, Y.; Petersen, J.N.; Buscheck, T.A.; Nitao, J.J. *Transport Porous Med.* **2002**, 49, 175–190.

21. Clement, T.P. *Water Resour. Res.* **2001**, *37*, 157–163.

22. Eykholt, G.R.; Li, L. *J. Contain. Hydrol.* **2000**, *46*, 163–185.

23. Bauer, P.; Attinger, S.; Kinzelbach, W. *J. Contain. Hydrol.* **2001**, *49*, 217–239.

24. Quezada C.R.; Clement, T.P.; Lee, K.K. *Adv. Water Resour.* **2004**, *27*, 508–521.

25. Nitao, J.J.; Buscheck, T.A. *Water Resour. Res.* **1991**, *27*, 2099–2112.

26. Berkowitz, B.; Zhou, J. *Water Resour. Res.* **1996**, *32*, 901–913.

27. Tang, D.H.; Frind, E.O.; Sudicky, E.A. *Water Resour. Res.* **1981**, *17*, 555–564.

28. Sudicky, E.A.; Frind, E.O. *Water Resour. Res.* **1982**, *18*, 1634–1642.
29. Sudicky, E.A.; Frind, E.O. *Water Resour. Res.* **1984**, *20*, 1021–1029.
30. Cormenzana, J. *Water Resour. Res.* **2000**, *36*, 1339–1346.
31. Sun, Y.; Buscheck, T.A. *J. Contam. Hydrol.* **2003**, *62-63*, 695–712.
32. Sun, Y.; Clement, T.P.; Petersen, J.N.; Hooker, B.S. A Modular Computer Model for Simulating Natural Attenuation of Chlorinated Organics in Saturated Groundwater Aquifers, Environmental Monitoring Tools Conference, Annapolis, MD, December, 1996.
33. Sun, Y.; Petersen, J.N.; Clement, T.P.; Hooker, B.S. *J. Contain. Hydrol.* **1998**, *31*, 147–162.
34. Rifai, S.H.; Bedient, P.B. *Water Resour. Res.* **1990**, *26*, 637–645.
35. Clement, T.P.; Sun, Y.; Hooker, B.S.; Petersen, J.N. *Ground Water Monit. Remediat.* **1998**, *18*, 79–92.
36. Clement, T.P.; Johnson, C.D.; Sun, Y.; Klecka, G.M.; Bartlett, C. *J. Contam. Hydrol.* **2000**, *42*, 113–140.
37. Nitao, J.J. *Reference Manual for the NUFT Flow and Transport Code, Version 2.0;* UCRL-MA-130651, Lawrence Livermore National Laboratory: Berkeley, CA, 1998.
38. Glascoe, L.G.; Wright, S.J.; Abriola, L.M. *J. Environ. Eng.* **1999**, *125*, 1093–1102.
39. Sun, N. *Water Resour. Res.* **2002**, *38*, 1029, 10.1029/2000 WR000198.
40. Xu, T.; Sonnenthal, E.L.; Spycher, N.; Pruess, K. *TOUGHREACT User's Guide. A Simulation Program for Non-isothermal Multiphase Reactive Geochemical Transport in Variable Saturated Geologic Media;* LBNL-55460, Lawrence Berkeley National Laboratory Report: Berkeley, CA, 2004.
41. Zheng, C.; Wang, P.P. *MT3DMS: A Modular Three-dimensional Multispecies Transport Model for Simulation of Advection, Dispersion, and Chemical Reactions of Contaminants in Groundwater Systems;* SERDP-99-1, US Army Engineer Research and Development Center: Vicksburg, MS, 1999.
42. Bateman, H. *Proc. Cambridge Phyil. Soc.* **1910**, *15*, 423.
43. Borden, R.C.; Bedient, P.B. *Water Resour. Res.* **1986**, *22*, 1973–1982.
44. Molz, F.J.; Widdowson, M.A.; Benefield, L.D. *Water Resour. Res.* **1986**, *22*, 1207–1216.
45. Kinzelback, W.; Schafer, W.; Herzer, J. *Water Resour. Res.* **1991**, *27*, 1123–1135.
46. Cirpka, O.A.; Frind, E.O.; Helmig, R. *J. Contam. Hydrol.* **1999**, *40*, 159–182.
47. Sun, Y.; Demir, Z.; Delorenzo, T.; Nitao, J.J. *Application of the NUFT Code for Subsurface Remediation by Bioventing;* UCRL-ID-137967, Lawrence Livermore National Laboratory: Berkeley, CA, 2000.

48. Waddill, D.W.; Widdowson, M.A. *J. Environ. Eng.* **1998**, *124,* 336–344.
49. Dortch, M.S.; McGrath, C.J.; Nitao, J.J.; Widdowson, MA.; Yabusaki, S. *Development of Simulators for in situ Remediation Evaluation, Design, and Operation. Final report;* ERDC/EL TR-01033, U.S. Army Corps of Engineers: Vicksburg, MS, 2001.
50. Radhakrishnan, K.; Hindmarsh, A.C. *Description and use of LSODE, the Livermore Solver for Ordinary Differential Equations;* UCRL-ID-113855, Lawrence Livermore National Laboratory: Berkeley, CA, 1993.
51. Gear, C.W. The automatic integration of ordinary differential equations, Morrell, A.J.H., ed., *Information Processing;* North-Holland Publishing Co.: NewYork, 1969; pp 187–193.
52. MathWorks *MATLAB High-performance Numeric Computation and Visualization Software;* Web site: www.mathworks.com: Natick, MA, 2000.
53. Yeh, G.T. *AT123D: Analytical Transient One-, Two-, and Three-Dimensional Simulation of Waste Transport in the Aquifer System;* ORNL-5602, Oak Ridge National Laboratory: CITY, TN, 1981; p 88.
54. Domenico, P.A. *J. Hydrology.* **1987**, *91,* 49–58.
55. Aziz, C.E.; Newell, C.J.; Gonzales, G.R.; Haas, P.E.; Clement, T.P.; Sun, Y. *BIOCHLOR–Natural Attenuation Decision Support System, Beta version 1.0, User's Manual;* Subsurface Protection and Remediation Division, National Risk Management Research: Laboratory: Ada, OK, 74820, 1999.
56. Clement T.P.; Truex, M.J.; Lee, P. *J Contam. Hydrol.* **2002**, *59,* 133–162.
57. van der Heijde, PLUME2D, www.mines.edu/igwmc/software/.
58. Wilson, J.L.; Miller, P.J. *J. Hydraul. Div.* **1978**, *104,* 503–514.
59. Valocci, A.J.; Werth, C.J.; Decker, J.J.; Hammond, G.; Zhou, P. *Interactive Model for Groundwater Flow and Solute Transport;* University of Illinois: Champaign, IL, 2003, www.cee.uiuc.edu/transport/.
60. Rifai, H.S.; Newell, C.J.; Gonzales, J.R.; Dendrou, S.; Kennedy, L.; Wilson, J. *BIOPLUME III Natural Attenuation Decision Support System version 1.0 User's Manual;* Brooks Air Force Base: San Antonio, TX, 1997.
61. Buscheck, T.E.; Alcantar, C.M. Regression Techniques and Analytical Solution to Demonstrate Intrinsic Bioremediation, In R.E. Hinchee et al., ed. *Intrinsic Bioremediation;* Battelle Press: Columbus, OH, 1995; pp 109–116.
62. Sun, Y.; Bear, J.; Petersen, J.N. *J Contam. Hydrol.* **2001**, *51,* 83–95.
63. Zhang, K.; Woodbury, A.D. *Adv. Water Resour.* **2002**, *25,* 705–721.
64. Zhang, G.X.; Zheng, Z.P.; Wan, J.M. *Water Resour. Res.* **2005**, *41,* W02018.

Chapter 10

Gaseous Transport Mechanisms in Unsaturated Systems: Estimation of Transport Parameters

Wa'il Y. Abu-El-Sha'r[1,2]

[1]Civil and Environmental Engineering Department, University of Central Florida, Orlando, FL 32816-2450
[2]Current address: Department of Civil Engineering, Jordan University of Science and Technology, Irbid, 22110, Jordan

Understanding gas transport in subsurface systems is necessary for the successful application of innovative technologies, such as soil vapor extraction (SVE), bioventing (BV), and in-situ air sparging (IAS), to remediate subsurface-contaminated hazardous waste sites. In general, gas transport in natural porous media occurs via several flux mechanisms including: viscous flow, free-molecule or Knudsen flow, continuum or ordinary diffusion (molecular and nonequimolar fluxes), surface flow or diffusion, and thermal flow. The relative importance of these mechanisms depends on factors related to the permeate gases (molecular weight, and viscosity) and the porous media (the type of porous media, porosity, moisture content, and bulk density). Three models are currently used to quantify the diffusive transport of gases in natural porous media, namely, Fick's first law, the Stefan-Maxwell equations, and the dusty gas model (DGM). The application of these models requires the quantification of many transport parameters, such as gas permeability coefficients, the effective binary diffusion coefficients, and Knudsen diffusion coefficients. These parameters are estimated from either laboratory or field measurements. Parameters determined from in-situ experiments tend to be much higher than the corresponding estimates from laboratory tests, while Knudsen diffusion coefficients have never been estimated in the field. These differences are attributed to the inclusion of macropores, fractures and heterogeneities in the field, and to the failure to consider all gas flux mechanisms that may have occurred. The DGM combines these mechanisms in a rigorous manner and can be applied to multicomponent gaseous systems under isobaric and non-isobaric conditions. Thus, it has been suggested as a framework for assessment and quantification of gas transport in subsurface systems and for the analysis of in-situ and laboratory tests to estimate the needed gas transport parameters.

Introduction

The successful application of innovative technologies, such as soil vapor extraction (SVE), bioventing (BV), and in-situ air sparging (IAS), to remediate subsurface-contaminated hazardous waste sites has served to focus increased attention on gas phase transport in subsurface systems. SVE involves applying a vacuum to the contaminated soil through extraction wells, which creates a negative pressure gradient that causes movement of vapors toward these wells (1). BV systems, on the other hand, deliver air from the atmosphere into the soil above the water table through injection wells placed in the ground where contamination exists. In a typical air sparging system, compressed air is delivered by a series of injection wells into the subsurface below any known point of contamination, and, as the air moves toward the surface, the contaminants are partitioned into the vapor phase. The contaminant-laden air rises toward the surface until it reaches the vadose zone, where an SVE system collects the vapors (1). The fundamental processes controlling these technologies are elaborate and involve airflow dynamics as well as contaminant transport, transfer, and transformation processes. A thorough understanding of these processes is necessary to evaluate the applicability and effectiveness of these technologies (1).

Gas transport in porous media occurs via one or more of the following mechanisms: viscous flow, Knudsen diffusion, bulk diffusion (non-equimolar and molecular diffusion), surface diffusion, and thermal diffusion. Different models exist to describe and combine some or all of these mechanisms. The use of these models requires certain transport parameters (diffusion and permeability coefficients) which may be estimated from correlations, if available, or by conducting specially designed experiments. These parameters depend on several factors representing the gases and the porous media under consideration.

This chapter focuses primarily on estimation of the parameters required for evaluating the different gaseous transport mechanisms in unsaturated zone by predictive models. The following issues are discussed:

1) Gas transport mechanisms in porous media,
2) Gas transport models,
3) Measurement or estimation of the various parameters required to quantify gas transport, and
4) In-situ estimation of gas transport parameters.

Although other fundamental processes, such as chemical mass transfer and biodegradation, are extremely important when evaluating the fate and transport of contaminants in the subsurface and designing of effective remediation technologies, these are not covered in this chapter and only certain aspects of gas transport in the subsurface are discussed.

Gas Transport Mechanisms in Porous Media

Gaseous systems are usually analyzed using the concept of small representative volume elements (RVEs). An RVE is macroscopic in size in that it contains enough molecules so that the properties of a gas are approximately constant within it and the volume is small enough to be treated as a point. Gas molecules in a given system frequently collide and constantly exchange momenta and kinetic energy with other gas molecules and with the particles of the porous medium (2, 3). These collisions are usually assumed binary (collision of two molecules only) and elastic (kinetic energy is conserved). According to the kinetic theory of gases, molecules of different gases at or very close to equilibrium have, on the average, the same kinetic energy, which is a function of the temperature of the system (2, 3). Hence, the lighter molecules tend to have larger velocities than the heavier ones.

The most important aspect of the theory of gas transport through porous media is that gas transport can be divided into different independent mechanisms, namely, viscous flow, free-molecule or Knudsen flow, continuum or ordinary diffusion (molecular and nonequimolar fluxes), surface flow or diffusion, and thermal flow. For a detailed discussion of these mechanisms, the reader is referred to various references (*2, 3, 4, 5, 6, 7*). In order to study these mechanisms, the porous media may be viewed as a collection of spherical particles (like giant molecules) distributed throughout the gaseous system. These particles are assumed fixed with respect to a given reference frame.

The following characteristic lengths may be defined to describe the transport mechanisms of gases in porous media: 1) the mean free path, λ, which is defined as the average distance between two consecutive molecular collisions, i.e., the average distance that a molecule travels before colliding with another molecule; 2) the average distance between two adjacent particles, λ_p; and 3) the particle radius, r_p. Note that λ_p and r_p are characteristics of the porous media and that the molecular size is negligible in comparison to these lengths. The gas transport mechanisms are discussed in details below.

Viscous Flux

If the mean free path, λ, is very small compared to the average distance between the adjacent particles, λ_p, and also very small compared to the particle radius, r_p (i.e., $\lambda \ll \lambda_p$ and $\lambda \ll r_p$), then the number of collisions between the gas molecules is very large compared to the collisions between the molecules and the walls of the particles. The molecules rebounding from the walls of the particles thus collide with the molecules flowing in the immediate vicinity of the particle, thus decreasing their average velocity. Because of this damping effect, a constant viscous flux results only if there is a pressure gradient to drive the molecules and to overcome the hindering effect of the rebounding molecules. This is that part of the flow in the continuum region that is caused by a pressure gradient.

Viscous flux depends on the coefficient of viscosity, which for gases is independent of pressure at constant temperature. Moreover, a mixture of gases behaves the same as a single gas because such bulk flow has no tendency to cause a mixture to separate into its components (no segregation of species). As a result, the viscous flux of any species in a mixture of gases is proportional to its mole fraction in the mixture.

Neglecting gravity effects, the viscous flux of ideal gases in the z direction in a porous medium is given by Darcy's law (*2*):

$$N^v = -\frac{k_g}{m}\frac{p}{RT}\frac{dp}{dz} \qquad (1)$$

where N^v is the total molar viscous flux in direction z; k_g is the effective gas permeability coefficient; μ is the viscosity; p is the pressure; R is the gas constant; and T is the temperature. For steady state conditions and small pressure gradients, eq 1 is often approximated in terms of average pressure $\bar{p}$ and a linear pressure gradient $\Delta p / \Delta z$.

Free-Molecule or Knudsen Flux

The Knudsen flux dominates when the mean free path of the molecules is extremely large compared to the pore radius ($\lambda \gg r_p$). This condition is sometimes expressed in terms of the Knudsen number, Kn ($\equiv \lambda / \lambda_p$), as Kn $\gg 1$. The Knudsen regime dominates when Kn ≥ 10 (8) since the number of molecule-molecule collisions is extremely small compared to the number of molecule-wall collisions. After molecule-wall collisions, the molecules rebounded or reemitted by the wall do not collide with other molecules, and, hence, the molecules of one species will not be affected by the presence of the other species. In other words, for a system in the Knudsen regime, there are as many independent fluxes present as there are species.

In a multicomponent system with Knudsen flux, each component may have a concentration gradient, and there will be a net flux of the molecules of that component. Then the net flux of gas i on both sides of a pore is proportional to the difference in the gas number densities at the two ends of that pore and can be estimated by (2):

$$N_i^k = -D_i^k \frac{\partial Ci}{\partial z} \qquad (2)$$

where N_i^k is the Knudsen molar flux of species I; D_i^k Knudsen diffusivity; and C_i is the molar concentration of species i. D_i^k is given by (2):

$$D_i^k = Q_p \left(\frac{RT}{M_i} \right)^{1/2} \qquad (3)$$

where Q_p is the obstruction factor for Knudsen diffusivity, also known as Knudsen radius, and is constant as a first degree of approximation; M_i is the molar mass of i. Another expression of D_i^k for a porous medium with a single pore size is given by (9):

$$D_i^k = 9.7 \times 10^3 \, \bar{r} \sqrt{\frac{T}{M_i}} \tag{4}$$

where $\bar{r}$ is the average pore radius.

Continuum or Ordinary Diffusion

Consider two gases (i, j) each contained in a closed system at the same temperature and pressure, connected by a tube filled with a porous medium. The first container holds gas i, which is lighter than the gas j in the other container. As the two gases leave their respective containers and enter the tube, the lighter molecules, i, according to the kinetic theory of gases, will have a larger speed than the heavier j molecules and will reach container j in less time, causing a pressure gradient to occur. If there are no walls (particles of the porous media), this pressure gradient will cause a nonsegregative viscous flux from container j towards i until this viscous flux is just sufficient to make the net flux zero. However, when there are walls, the pressure gradient will dissipate with loss of momentum, and the system will reach a quasi-steady-state where the net molecular flux is zero, but the pressure gradient is not.

This pressure gradient can be eliminated by external means, but the net molecular flux will then no longer be zero. Typically, diffusive fluxes are assumed equal and opposite in a binary system, but, for this binary system with no external applied pressure gradient, the net flux is not zero. This indicates that a non-segregative diffusive flux has developed as a result of the difference in concentration (non-equimolarity) of species and is called nonequimolar flux, also known as diffusion slip flux. Thus, in a system with walls, the total diffusive flux of a given species consists of two components-diffusive flux (segregative) and non-equimolar flux (non-segregative).

For the system described above, the total molar diffusive flux for gas i is given by (2):

$$N_i^D = J_{iM} + x_i \sum_{j=1}^{v} N_j^D \tag{5}$$

where N_i^D, N_j^D are the total molar diffusive fluxes of species i, and j, respectively; J_{iM} is the molar diffusive flux of species i; x_i is the molar fraction of species i; and v is the number of gas components. Eq 5 indicates that the total molar diffusive flux of species i consists of two components, namely the molar diffusive flux, J_{iM}, and the non-equimolar flux, $x_i \sum_{j=1}^{v} N_j^D$. The total molar diffusive fluxes for the different gas components of a gas mixture are related by Graham's law (3):

$$\sum_i M_i^{1/2} N_i^D = 0 \tag{6}$$

where M_i is the molecular weight of gas i.

Surface Diffusion

Surface diffusion occurs when the molecules of the diffusing gas are adsorbed to the low energy sites at the surfaces of the soil particles and then transferred along the surface by hopping from one adsorption site to another until they gain enough energy to escape from the soils surface and return to the gaseous phase. Each of the species transported along the surface behave independently of the others and, thus, its flux is only proportional to its own surface concentration gradient, which, in turn, is proportional to its gas-phase concentration (3).

The surface diffusive flux, at low surface coverage, is usually modeled by Fick's law of diffusion with the concentration gradients refer to the surface concentration gradients (eq 7). All complexities of the porous medium geometry, surface structure, adsorption equilibrium, etc. are contained in the surface diffusion coefficients,

$$J_{iS} = -D_{iS} \nabla n_i \tag{7}$$

where J_{iS} is the surface flux referred to apparent unit cross-sectional area of the medium; D_{iS} is a surface diffusion coefficient; and ∇n_i is the molecular surface concentration gradient.

This mode of transport has been neglected in many studies that dealt with gaseous transport in natural soils under steady state conditions because of the common belief that the number of molecules that will adsorb to the soil surfaces is equal to the number of those leaving these surfaces to the gaseous phase.

Thermal diffusion

Thermal diffusion is the relative motion of the gaseous species caused mainly by temperature gradient. The thermal diffusion flux for gas i, N_i^{Th}, in a gaseous mixture is given by:

$$N_i^{Th} = C \sum_{j=1}^{n} x_i x_j \left(\frac{D_i^T}{\rho_i} - \frac{D_j^T}{\rho_j} \right) \rho_m \nabla \ln T \tag{8}$$

where D_i^T, D_j^T are thermal diffusion factors of components i and j, respectively; n is number of gas components; ρ_i, ρ_j, and ρ_m are the mass densities of gases i, j, and the gaseous mixture, respectively; and T is the temperature. It has been theoretically demonstrated that, for temperature differences less than 200°C, the thermal diffusive flux is insignificant and may be ignored (10).

Gas Transport Models

Three distinct models are currently used to quantify diffusive transport of gases in natural porous media, namely: Fick's first law, the Stefan-Maxwell equations, and the dusty gas model. These models are briefly reviewed in the following subsections.

Fick's First Law of Diffusion

Fick's first law of diffusion is generally used to predict molecular diffusion of a binary gas system into porous media under isobaric conditions. For multicomponent systems, a binary analysis is performed with all gas components, other than the gas for which the prediction is needed, assumed as one component. For non-isobaric conditions, the diffusive and the viscous fluxes are usually assumed to be independent of each other and thus the total flux is obtained by adding these two fluxes. The diffusive flux given by Fick's law of diffusion when written in the molar form is given by (11, 12):

$$\left(N_i^D \right)_F = -D_{ij}^e C \nabla x_i \tag{9}$$

where $\left(N_i^D \right)_F$ is the Fickian molar diffusive flux of component i; D_{ij}^e is the effective binary diffusion coefficient of gas components i and j (given by eq 10 below); C is the total molar gas concentration; and ∇x_i is the molar fraction gradient for species i. The effective binary diffusion coefficient of components i and j can be written as:

$$D_{ij}^e = Q_m \, D_{ij} \qquad (10)$$

where Q_m is the diffusibility factor, also known as the obstruction factor, and is only a function of the porous medium; and D_{ij} is the free binary diffusion coefficient of gases i and j.

Eq 9 has the same form as Fick's law for a liquid provided that the total concentration of the gas system is constant, i.e., for an isobaric system. Despite the fact that Fick's law of diffusion is widely used in soil science and environmental engineering to study gaseous diffusion, it must be remembered that, when applied to gases, it is a primarily an empirical relation borrowed from studies involving solutes and shown to agree well with some observations of gaseous diffusional processes (*13*).

For non-isobaric systems, Fick's law given by equation (*9*) have been assumed to hold and superimposed with Darcy's law (eq 1).

Dusty Gas Model (DGM)

The dusty gas model (DGM) combines the different gas transport mechanisms in a rigorous manner and can be applied to multicomponent systems, both isobaric and non-isobaric. Here, a full Chapman-Enskog kinetic theory treatment is given for a gas mixture in which the porous media is considered as one component of the mixture (*2, 3, 4*). This approach enables all the elaborate results of modern kinetic theory, constitutive equations, to be used without having to repeat the Chapman-Enskog treatment.

The physical picture behind the model is that of a dusty gas, in which the dust particles constitute the porous media. The basic working assumptions of the DGM are 1) the suspended particles are spherical, can be treated as a component of the gas mixture, are motionless and uniformly distributed, are very much larger and heavier than the gas molecules, and are acted upon by an external force that keeps them at rest even though a pressure gradient may exist in the system and 2) no external forces act on the gas molecules.

The constitutive forms of the DGM equations for v gas components for isobaric isothermal conditions are given by (*2*):

$$\sum_{i=j,j\neq 1}^{v} \frac{x_i\left(N_j^D\right)_{DGM} - x_j\left(N_i^D\right)_{DGM}}{D_{ij}^e} - \frac{\left(N_i^D\right)_{DGM}}{D_i^k}$$

(11)

$$= \frac{1}{RT}\nabla p_i + n'\sum_{j=1}^{v} x_i x_j \alpha_{ij} \nabla \ln T$$

where x_i and x_j are the molar fraction of gas component i and j, respectively; $\left(N_i^D\right)_{DGM}$ and $\left(N_j^D\right)_{DGM}$ are the total molar diffusive fluxes of components i and j given by the DGM; D_i^k and D_j^k are the Knudsen diffusion coefficient of species i and j, respectively; p_i is the component pressure of gas i; R is the gas constant; T is the temperature of the system; n is the gas and particle density; α_{ij} is the generalized thermal diffusivity; and v is the number of gas components.

The summation term on the left-hand side of eq 11 is the momentum lost through molecule-molecule collisions with species other than i (but not the particles). The second term on the left hand-side of eq 11 is the momentum lost by species i through molecule-particle collisions. The first term on the right-hand side represents the concentration gradient contribution to diffusion of species i, and the second term represents the thermal gradient effect on diffusion, which may be neglected for isothermal systems.

The DGM theory for flow of gases in man-made porous media has been experimentally tested under isothermal and non-isothermal conditions (3). The DGM theory was also experimentally verified for gas transport in natural soils under-isobaric condition (7). Few studies reported using this model in the environmental engineering or hydrology literature (5, 6, 8, 14, 15, 16, 17, 18, 19, 20).

Stefan - Maxwell Equations

Stefan-Maxwell (SM) equations are commonly used by chemical engineers to study multicomponent diffusion. If the second term of the left-hand side of eq 11 is negligible, i.e., in the molecular regime, eq 11 reduces to Stefan-Maxwell equations:

$$\sum_{i=1,j\neq 1}^{v} \frac{x_i\left(N_j^D\right)_{SM} - x_j\left(N_i^D\right)_{SM}}{D_{ij}^e} = \frac{p}{RT}\nabla x_i$$

(12)

where $\left(N_i^D \right)_{SM}$ and $\left(N_j^D \right)_{SM}$ are the total molar diffusive fluxes of components i and j, respectively.

The same equations can be written in a form similar to Fick's law, where the diffusion coefficients are functions of the composition of component gases in addition to the porous medium.

Choice of Model

Of the three modeling approaches described previously, Fick's first law of diffusion has been used most extensively to model the diffusive transport of gases in natural porous media (soils). This relation was originally verified experimentally for liquid phase transport but has often been extended to predict gas diffusion without recognition of the differences between liquid and gas properties (2). A number of recent studies in the hydrology literature have explored the adequacy of Fick's first law of diffusion to model transport in the gaseous phase and the importance of other mechanisms of gas transport, such as nonequimolar, Knudsen, and viscous fluxes (5, 6, 7, 12, 15, 16, 17, 18, 19, 20, 21, 22). Results of the above studies reveal that a complete understanding of gas transport in the unsaturated zone will require a multicomponent analysis and consideration of a number of gas flux mechanisms.

Neither Fick's first law of diffusion nor Stefan-Maxwell equations can adequately model conditions where Knudsen diffusion is important, and the DGM is recommended. Under non-isobaric conditions, viscous fluxes may be superimposed with the total diffusive fluxes to estimate the total flux of a particular gas phase component. If Fick's law of diffusion or the Stefan-Maxwell equations, originally developed for isobaric conditions, are used to analyze such systems, an assumption regarding the additivity of the diffusive and viscous fluxes is needed. However, if the DGM is used, the simple addition of viscous and diffusive fluxes appears to have been first proposed using electric analog arguments (23). Nevertheless, the simple addition of the diffusive flux (given by any of these three models) and viscous flux (given by Darcy's law) has been found to underestimate the actual gaseous fluxes, and a careful examination of the theory is needed (7).

In most applications to natural soils, surface diffusion has been always neglected. For non-isothermal systems, the thermal diffusion flux is added to the total flux. When the temperature gradient is less than 200°C, the thermal diffusive flux may be neglected (10).

Table I below summarizes the suitable conditions under which a given gas transport flux mechanism may significantly affect gas transport predictions.

Measurements or Estimation of the Various Parameters
Required to Quantify Gas Transport

Numerous experimental studies have been conducted to estimate the transport parameters needed by transport models to assess and quantify gas transport in natural porous media. These include: close, semi-open, or open experiments; single, binary, or multicomponent gaseous systems; dry or wet conditions; natural or man-made porous media; and gas permeability coefficients, Knudsen radius, or diffusibility determination. The following subsections briefly describe the basics of the experimental methods used and summarize the relevant studies reported.

Table I. Gas Transport Flux Mechanisms To Be Considered Under Different Conditions

Flux mechanisms	Suitable conditions
Molecular	All types of natural porous media, all gaseous systems, and any environmental conditions
Non-equimolar	All gaseous systems except for binary systems with equimolar pair of gaseous
Knudsen	Very fine material with permeability less than 10^{-10}cm^2, fine soils with saturation above 65%, or under extremely low total pressures
Thermal	For temperature differences greater that 200 °C
Surface	Transient conditions and whenever there is a potential physical adsorption of flowing gases onto the soil matrix

Experimental Systems

Experimental systems that have been or may be used for gas transport parameters estimation include closed, semi-open and open systems as described below.

Closed System

In this system, the component gases are not added or taken from the system and are usually represented by two containers each containing one or more of the component gases. These containers are connected to each other through a capillary of porous media. In this system, the total diffusive fluxes of the two component gases are equal and opposite in direction, and the total flux is given by eq 5.

Semi-Open System

This system is usually thought of in terms of Stefan's tube, which is simply a tube that contains the material to be diffused in its liquid form at the tube's bottom. The top of the tube can be left open to air when measuring free diffusivity, or can be plugged with the porous media of interest when estimating the effective diffusion coefficients. Analysis of this system can be carried out by either Fick's law, the SM equation, or the DGM equations.

Open System

In this system, the component gases are allowed to flow passing the edges of the porous soil sample. This is the most general system and has an advantage in that the pressure gradient and absolute pressures can be regulated by controlling the flow rate of the component gases. The analysis of this system is usually carried out by the DGM.

Characterization of the Porous Media

A full characterization of any porous medium (under isothermal conditions) can be attained by determining three transport parameters, namely, the coefficient of permeability, the Knudsen radius (required to estimate the Knudsen diffusion coefficient for all gas components present in the system), and the diffusibility (needed to estimate effective binary diffusion coefficients for all gas pairs in the system). The number of parameters needed depends on the model used and the modes of transport considered for the particular situation of interest. While one parameter may be enough to study a diffusive transport by Fick's law, three parameters are needed to study gas transport by DGM. The following subsections present a brief summary of the relevant studies involving each transport parameter.

Knudsen Diffusion coefficients

Knudsen diffusion coefficients may be estimated from specially designed experiments involving mainly single-gas experiments as described below:

Single-Gas Experiments

The only experiments possible using single-component gaseous systems involve forcing the gas through a porous medium by a pressure difference (permeability experiments). These experiments may be used to measure the Knudsen diffusion coefficient as well as the coefficient of permeability of the porous medium. Traditional permeability experiments using liquids are based on the assumption that, as long as the volumetric flow rate is proportional to the pressure gradient, the coefficient of permeability is a property of the porous media only and is independent of the fluid used in its determination. However, when Klinkenberg (*24*) studied the permeability of rock core samples and glass filters (materials of permeability less than 10^{-10} cm^2) to various liquids and different gases (air, hydrogen and carbon dioxide) at different pressures, the coefficient of permeability, estimated from experiments conducted with gases, was found out to be dependent on the nature of the gas and is approximately a linear function of the reciprocal mean pressure. The values of the gas permeability were greater than those of the liquid permeability and, when extrapolated to infinite pressure, gave values equal to those of the liquid permeability, which is only a function of the porous media. Klinkenberg explained this effect by considering the phenomenon of slip, which is related to the mean free paths of the gas molecules and, thus, Knudsen diffusion. These observations are sometimes referred to as the "Klinkenberg effect."

For a pure component (i) in the gas phase, the DGM equation (eq 11) reduces to the following form (*3*):

$$\left(N_i^D\right)_{DGM} = -\left(\frac{D_i^k \mu}{\overline{P}} + k\right)\frac{\overline{P}}{\mu RT}\nabla P \tag{13}$$

A comparison of eq 13 with the Darcy's law expression given by eq 1 suggests that the total molar flux in this system may be thought of as a viscous flux with the intrinsic permeability replaced by an apparent or gas phase permeability, k_g:

$$N_i^T = -\frac{\overline{P}}{RT}\frac{k_g}{\mu}\nabla P \tag{14}$$

where

$$k_g = \frac{D_i^k \mu}{\bar{P}} + k \tag{15}$$

Eq 14 indicates that, if N_i^T is plotted as a function of ∇P, the slope of the line that best fits the data and passes through the origin can be used to calculate k_g. Consistent with eq 15, it is well known that the values of k_g obtained in this manner for different gases are greater than the equivalent liquid or intrinsic permeability, k, and depend upon the values of the mean pressure, $\bar{P}$, of the gas (*2, 3, 7, 17, 24*). As $\bar{P}$ increases, the gas permeability decreases until it approaches the liquid permeability, k. At low pressures the first term in eq 15 gives rise to what is known in the literature as "slip flux" (*24*).

Eq 13 suggests that a plot of $N_i^T RT/\nabla P$ versus $\bar{P}$ should yield a straight line with an intercept of D_i^k and a slope of k/μ. Such plots have been reported to estimate k and D_i^k for each soil sample.

Summary of Measurements of Knudsen Diffusion Coefficients

There have been very few laboratory measurements of the Knudsen diffusion coefficient for unconsolidated soils. Stronestrom and Rubin (*25*) examined the relationship between saturation and Klinkenberg factor for two sandy soils. Abu-El-Sha'r (*7*) measured the Knudsen diffusion coefficient for sea sand, Ottawa sand, kaolinite clay and several sand mixtures, and kaolinite sea sand mixtures. The effects of low to intermediate levels of water saturation on the Knudsen diffusion coefficient were also investigated. Correlations were developed for the Knudsen radius as a function of the characteristic length of the soil, which is defined as the square root of the true permeability (*17*), and for the effect of moisture content on the Knudsen coefficient (*7*). Reinecke and Sleep (*26*) measured the Knudsen diffusion coefficient at different levels of water saturation for unconsolidated porous medium consisting of silt-sized particles. At high levels of water saturations (>64%), Knudsen diffusion measurements indicated that Knudsen diffusion has significant implications for the prediction of organic vapor transport in partially saturated silts and in less permeable soils. An expression was developed for the influence of water saturation on Knudsen diffusion coefficients using Brooks-Corey capillary pressure saturation relationship. They concluded that Knudsen diffusion may play an important part in gas transport in fine porous media (permeability less than 10^{-10} cm^2), but, for sandy materials, molecular diffusion is expected to dominate. Table II summarizes the available expressions for estimation of Knudsen diffusion coefficients in natural unconsolidated soils.

Table II. Correlations for Knudsen Diffusion Coefficient in Soils

Correlation	Reference	Soils Tested	Comments
$Q_p = 0.00182$ $- 8.199\,L_c$ $+ 14945\,L_c^2$	Abu-El-Sha'r- Abriola (17)	Sea sand, Ottawa sand, sand mixtures, kaolinite clay, and kaolinite clay –sea sand mixtures.	Open system, Helium, dry samples (Note: L_c is the characteristic length of the porous media.)
$Q_p = -0.049$ $+0.2806\,n_a -$ $0.3485 n_a^2$	Abu-El-Sha'r (7)	Sea sand	Volumetric moisture content below 20% (Note: n_a is the air filled porosity.)
$D_{air}^K = 2.69$ $\times 10^6 k^{0.764}$	Reinecke- Sleep (26)	Silt-sized particles	Wide range of saturation, correlation included all previous measurements

Binary Diffusion Coefficient

Binary Gas Experiments

These experiments determine the obstruction factor (also known as diffusibility) by which the free diffusivity of any pair of gases must be adjusted to take into account the effect of the porous medium tested. The obstruction factor is assumed to be solely a function of the porous medium and is usually expressed as a function of air-filled porosity and tortuosity. Although binary and multicomponent experiments have been reported to measure the binary diffusion coefficients, experiments conducted with a pair of gases under isobaric and isothermal conditions are the most common and reliable. For a binary gaseous system, under isothermal (T = constant) and isobaric (C = constant) conditions, the DGM (eq 11) yields an explicit form for the diffusive flux of component i (2),

$$\left(N_i^T\right)_{DGM} = -\left\{\frac{1}{D_i^k} + \frac{1-\left(1-\left(M_i/M_j\right)^{1/2} x_i\right)}{D_{ij}^e}\right\}^{-1} C\nabla x_i \qquad (16)$$

If an equimolar pair $(M_i\, M_j)$ of gases are used to carry out the experiment, eq 16 can be simplified as:

$$\left(N_i^T\right)_{DGM} = -\left\{\frac{1}{D_i^k} + \frac{1}{D_{ij}^e}\right\}^{-1} C\nabla x_i \qquad (17)$$

Eq 17 combined with the ideal gas law and the Knudsen diffusivity expression (eq 3) gives the following form:

$$\frac{\nabla x_i \overline{P} D_{ij}}{\left(N_i^D\right)_{DGM} RT} = \frac{1}{Q_m} + \frac{1}{Q_p} \frac{\overline{P} D_{ij}}{\left(RT/M_i\right)^{1/2}} \frac{1}{\overline{P}} \qquad (18)$$

If the left-hand side of eq 18 is plotted as a function of $1/\overline{P}$, a straight line is obtained, where Q_m can be computed from the intercept and Q_p can be computed from the slope.

For coarse material, the Knudsen diffusivity is large; diffusion is in the molecular regime; and the second term on the right-hand side of eq 18 is negligible, yielding

$$\frac{\nabla x_i \overline{P} D_{ij}}{\left(N_i^D\right)_{DGM} RT} = \frac{1}{Q_m} \qquad (19)$$

If Fick's law of diffusion is used (eq 9) to analyze such experiments, the measured diffusion coefficient will include both the molecular diffusion and the non-equimolar fluxes (if a non-equimolar pair of gases is used) and thus does not depend on the porous media alone.

Summary of Diffusibility Measurements

In the hydrology and environmental engineering literature, the reported measurements of transport parameters in natural soils have centered on the effective binary diffusion coefficients, also known as diffusibilities (*7, 8, 17, 27, 28, 29, 30*). Limited measurements have been reported that consider or account for flux mechanisms other than the molecular diffusion modeled by Fick's first law (*7, 8, 16, 17, 26*). In addition, all previous diffusion measurements, except for the measurements reported in (*7, 17*), were based on experiments conducted using pairs of gases with different molecular weights, i.e., all studies overlooked the potential importance of non-equimolar flux mechanisms in the evaluation of diffusibilities. Table III summarizes the estimated diffusibility correlations that have been reported in the literature and the applicable porosity range of each correlation as provided in reference (*7*).

In the soil science, hydrology, and environmental literature many experimental studies involving transport of organic compounds vapors in soil materials have been reported. The relevant studies and their major findings are chronologically presented in Table IV.

Moisture content can have important effects on the diffusion of gases in soils. These effects may be due to the reduction in the effective cross-sectional area available for diffusion, the blocking of the available sites for adsorption, and the solubilization of the gas in the water surrounding the soil particles. It also alters the air-filled porosity, the flow paths of gases, and the role of Knudsen diffusion. On the other hand, soil bulk density has an exponential effect on gas flux through soil (*38, 39, 40*). However, there is a limit to the maximum bulk density obtained, and there will always be a finite amount of open space for gas diffusion to take place.

Other properties affecting gas transport in subsurface systems include porosity, air-filled porosity, pore size distribution, type of soil (organic content), and temperature. These properties may influence the values of the transport parameters and control the dominant gas transport mechanism.

In-situ Estimation of Gas Transport Parameters

Three main sets of gas transport parameters are needed for fully understanding gas transport in natural subsurface systems under isothermal conditions. These include gas permeability coefficients, molecular diffusion coefficients, and Knudsen diffusion coefficients. Reported in-situ measurements focused on determination of gas permeability coefficients and, to a lesser degree, on molecular diffusion coefficients. No estimation of Knudsen diffusion coefficients at field conditions has been identified yet. In-situ gas permeability coefficients are estimated from analysis of pneumatic tests, atmospheric pumping data, or measurements by air mini permeameters (41).

Pneumatic tests involve pumping or extracting air from a certain well (vertical or horizontal) while monitoring gas pressure in different gas ports installed at different surrounding locations and different depths. The tests are usually conducted in a way to minimize water redistribution during the test by controlling the injection or extraction rates. Both the initial transient phase or the steady-state part of the test can be analyzed to obtain gas permeability from pressure data. Many solutions exist based on the method of solutions and the boundary conditions used. These solutions generally assume radial flow and no flow boundary at the water table level. Many solutions assume ideal gas law conditions and neglect the Klinkenberg effects (related to Knudsen) (42, 43). A summary of the available methods of analyzing the pneumatic tests data is provided by reference (41).

The temporal response of soil gas pressure at different depths to atmospheric pressure fluctuations can be used to estimate the vertical air permeability between ground surface and the corresponding depth (44, 45). Gas pressures are monitored at different depths in the unsaturated zone by using differential pressure transducers placed at the surface, with one port of the transducer connected to a gas port (screened intervals) at a certain depth and the other left open to the atmosphere. Data analysis starts by expressing the variations in atmospheric pressure as time harmonic functions. Then the attenuations of surface waves at different depths in the unsaturated zone are determined and used to estimate the vertical air permeability. A solved example for estimating gas conductivity using subsurface attenuation of barometric pressure fluctuations is presented in (41).

Gas permeability at a localized scale has been measured by minipermeameters, devices that inject pressurized nitrogen gas at a constant pressure through a tip pressed against the measurement surface. Both the steady-state flow rate and the injection pressure at the tip are measured and compared to calibration curves of pressure and flow rates for materials of known permeability. Gas permeability values obtained from field data are generally greater than the corresponding values obtained from laboratory experiments. In some cases, the field values exceed the laboratory values by orders of magnitude (44, 46). These differences are attributed to the increase in scale from laboratory to field measurements and the inclusion of macropores, fractures and heterogeneities in the field. Also, field permeability measurements in low permeability media or aquifer material with high saturation include the effects of Knudsen diffusion, which are not usually considered in the analysis.

In-situ effective binary diffusion coefficients have been estimated from field experiments (30, 47, 48, 49, 50, 51, 52). For shallow depths, near the ground surface, a tube is inserted into soil surface and covered with a gas

Table III. Summary of Reported Diffusibility Correlations (7)

Reference	Diffusibility, Q_m	Comments
Buckingham (1904)	$1/(1/n)^2$	CO_2 in air, soils, steady state
Bartell and Osterhof (1928)	$n/(\pi/2)$	---
Brugeman (1935)	$(n)^{2/3}$	Theoretical, not for mono-sized spheres
Penman (1940)	$2/3\ (n)$	Carbon disulfide and acetone vapors in air, granular soils, steady State
Carman (1956)	$n/[\cos45]^2$	---
Marshall (1958)	$(n_a)^{3/2}$	Pseudo-theoretical
Millington (1959)	$(n_a)^{4/3}$	Pseudo-theoretical
Millington and Quirk (1961)	$n_a^{10/3}/n^2$	Theoretical approach
Masamune and Smith (1962)	$n/[1+3(1+n)]^2$	---
Dumanski	$n/[1-(1-n)^{0.67}]^2$	---
Maxwell	$n/[(3-n)/2]^2$	Theoretical, ordered packing
Weissberg (1963)	$n/[1-0.5*\ln(n)]^2$	Theoretical, overlapping spheres
Johnson and Stewart (1965)	$n/4$	---
Sallerfield and Cadle (1968)	$n/7 < Qm < n/3$	Catalysts
Currie (1970)	$(n)^{3/2}$	H_2 into air, dry sands
Currie (1970)	$(n_a/n)^4(n)^{3/2}$	H_2 into air, wet sands
Millington and Shearer (1970)	$(1-S_w)^2\ (n-\theta)^{2x}$	Theoretical
Neale and Nader (1973)	$2n/(3-n)$	Theoretical, random homogeneous isotropic packing
Lai et al (1976)	$(n_a)^{5/3}$	O_2 into air, undisturbed natural soil.
Cunninghan and Williams (1983)	$n/3$	Catalysts
Abu-El-Sha'r and Abriola (1997)	$0.435n$	CO and N_2 into sea sand, Ottawa sand, kaolinite clay, sand mixtures, and sea sand-kaolinite mixtures

Note: n is the porosity, and n_a is the air filled porosity.

Table IV. Summary of Experimental Studies Involving Organic Gases/Vapors

Reference	Vapors/gases	Description	Findings
Penman (1940) Source: (27)	Carbon disulphide and acetone	Semi-open system; steady and unsteady state conditions: diffusive fluxes only; granular soils	Diffusion coefficients in porous media are related to diffusivity coefficient in air, the air-filled porosity, and the tortuosity of air-filled pores
Alzaydi (1975) Source: (8)	Gas couples of N_2, CH_4, CO_2, O_2, and air	Very long packed columns; unsteady conditions; combined flow; Ottawa sand and kaolinite clay	Measured diffusion coefficients for all gas couples used; developed a relation between tortuosity and pore radius.
Farmer et al (1980) Source: (31)	Hexachloro-benzene (HCB)	Semi-open system; Silty clays form a municipal landfill; Steady state; Diffusive fluxes: Assumed binary with Fick's law	Millington and Quirk (1961) correlation can be used to predict the diffusion coefficients when the porosity of the porous media is greater than 0.30. Also, an increase of 13.5% in the soil air-filled porosity caused an exponential increase (45%) in the apparent HCB diffusion coefficient.
Thomson (1985) Source: (32)	Trichlorofluoro-methane, Methylene chloride, chloroform, 1,1,1, trichloroethane, carbon tetrachloride, trichloroethene, and perchloroethene	Semi-open system; gravel (bottom 50cm) and washed sand (top 120 cm); unsteady state Analysis: Fick's second law	Measured diffusion coefficients and retardation factors for all the vapors used into air.

Continued on next page.

Table IV. *Continued.*

Reference	Vapors/gases	Description	Findings
Baehr and Bruell (1990) Source: (*21*)	Benzene, hexane, and isooctane vapors	Semi-open system; sand from alluvial deposits; steady state conditions; both Stefan - Maxwell and Fick's law were used	Determined tortuosity factors based on both models. Tortuosity is overestimated when Fickian diffusion is assumed.
Wakoh and Hirano (1991) Source: (*33*)	Propane	Field measurements of leaked propane; predictive model base on convection diffusion (only Molecular diffusion); compared predictions with measurements	In order for the model to predict the measured concentrations, diffusion coefficients must be much higher than the effective molecular diffusion coefficients commonly estimated
Fuentes et al. (1991) Source: (*34*)	TCE and orthoxylene	Semi-open system; undisturbed tuff; Fick's law of diffusion; Moist samples (1 to 3%, and 12to 15% (by mass))	The flux of the organics was reduced by an order of magnitude when the moisture content was Raised from 1 - 3%, to 12 - 15% (by mass)).

Table IV. *Continued.*

Reference	Vapors/gases	Description	Findings
Johnson and Perrott (1991) Source: (35)	Gasoline hydrocarbons, carbon monoxide, methane and oxygen	Included laboratory, field and modeling based on Fick's second law for one year; fine silty loam; high water content; diffusion coefficients estimated from steady state experiments using a semi-open system.	The effective diffusion coefficients for wet samples are much smaller than those of dry soils; the tortuosity factors (obstruction factor) calculated from the measure diffusion coefficients were similar to those predicted by Millington (1959).
Hutter et al (1992) Source: (36)	TCE	Semi-open system; large soil column; moist sand, silt, and clay mixture (% weight: sand 73.3%; gravel 10%; silt 7.5%; clay 9.2%). Air-filled porosity ranged from 0.29 to 0.43 (final moisture content around 9%). Fick's law	Measured the diffusion coefficients for TCE for different porosities of the soil sample.
Moldrup et al 2000 Source: (37)	Oxygen	126 undisturbed soil samples at two values soil water potential including -100 cm water; Loamy sand; Fick's law	Developed a correlation for effective diffusion coefficient as a function of soil water characteristics

chamber that contains the gas to be used for measurement. Then the gas composition in the chamber is determined at different times. Knowing the air-filled porosity of the soil, the initial gaseous composition in both the soil and the chamber is then used to estimate the diffusibility needed for calculating the effective binary diffusion coefficient. The models used in analyzing the results are usually based on Fick's law of diffusion. For other depths, a permeation device is used to release a tracer gas along with continuous monitoring of its concentration for a certain period of time (47). The data is then used to estimate the diffusibility factors based on either analytic or numerical methods.

Summary

This chapter presents a thorough review of gas transport mechanisms in unsaturated systems, commonly used gas transport models, and the laboratory and in-situ methods used to estimate the parameters needed for these models. Gas transport in natural porous media can be divided into several modes of transport or flux mechanisms, such as viscous flow, free-molecule or Knudsen flow, continuum or ordinary diffusion (molecular and nonequimolar fluxes), surface flow or diffusion, and thermal flow. These are commonly quantified by models, such as Fick's first law, the Stefan-Maxwell equations, and the dusty gas model (DGM). The use of these models, however, requires accurate estimates of several transport parameters, including gas permeability, effective binary diffusion coefficients, and Knudsen diffusion coefficient. Both laboratory and in-situ methods have been used to provide the needed parameters. A comprehensive survey of the relevant literature reveals that Knudsen diffusion coefficients have never been estimated in the field; the estimates of permeability coefficients and effective diffusion coefficients from field measurements are significantly larger than those obtained from laboratory experiments; and many reported studies ignored one or more of the gas flux mechanisms that may have occurred. Consequently, the dusty gas model (DGM) has been suggested for the analysis of gaseous systems because it combines all gas flux mechanisms in a rigorous manner and can be applied to multicomponent gaseous systems under isobaric and non-isobaric conditions.

References

1. Sharma, H.D.; Reddy, K.R. *Geoenvironmental Engineering: Site Remediation, Waste Containment, and Emerging Waste Management Technologies;* John Wiley & Sons, Inc.: Hoboken, NJ, 2004.

199

2. Cunnigham, R. E.; Williams R.J.J. *Djffusion in Gases and Porous Media;* Plenum Press: New York, NY, 1980.
3. Mason, E. A.; Malinauskas A.P. *Gas Transport in Porous Media; the Dusty Gas Model;* Elsevier Scientific Publishing Company: New York, NY, 1983.
4. Jackson, R. *Transport in Porous Catalysts;* Elsevier Scientific Publishing Company: New York, NY, 1977.
5. Thorstenson, D. C.; Pollock, D.W. *Rev. Geophys.* **1989**, *27(1),* 61-78.
6. Thorstenson, D. C.; Pollock, D.W.. *WRR.* **1989**, *25 (3),* 407-409.
7. Abu-El-Sha'r, W. Y. Ph.D. thesis, The University of Michigan, Ann Arbor, MI, 1993.
8. Alzaydi, A. A. Ph.D. thesis, The Ohio State University, Columbus, OH, 1975.
9. Geankoplis, C. J., *Mass Transport Phenomena;* Holt, Rinehart and Winston, Inc.: New York, NY, 1972.
10. Al-ananbeh, A. M.Sc. thesis, Jordan University of Science and Technology, Irbid, Jordan, 2003.
11. Bird, R.B.; Stewart, W.E.; Lightfoot, E.N. *Transport Phenomena;* John Wiley & Sons, Inc.: New York, NY, 1960.
12. Jaynes, D. B.; Rogowski A.S. *Soil Sci. Soc. Amer. J.* **1983**, *47,* 425-430.
13. Kirkham, D.; Powers, W.L. *Advanced Soil Physics;* Wiley Interscience.: New York, NY, 1972.
14. Alzaydi, A. A.; Moor, C. A.; Rai, I. S. *AIChE Journal.* **1978**, *24 (1), 35-42.*
15. Abriola, L. M.; Fen C. S.; Reeves H.W. In *Subsurface Contamination by Immiscible Fluids;* Weyer, K.U., Ed.; IAH Conference Proc., Calgary, Canada, 1990., Balkema, 1992; Rotterdam, Netherlands, pp 195-202.
16. Voudrias, E. A.; Li, C. *Chemophere.* **1992**, *25(5),* 651-663.
17. Abu-El-Sha'r, W. Y.; Abriola, L. M. *WRR.* **1997**, *33(4),* 505-516.
18. Sleep, B.E. *Advances in Water Resources.* **1998**, *22(3),* 247-256.
19. Fen, C.-S. *Geologica,* **2002**, *46(2-3),* 436-438.
20. Fen, C.-S.; Abriola, L.M. *Advances in Water Resources.* **2004**, *27(10),* 1005- 1016.
21. Baher, A. L.; Bruell, C.J. *WRR.* **1990**, *26(6),* 1155-1163.
22. Allawi, Z. M.; *AJChE Journal.* **1987**, *33(5),* 766-755.
23. Mason, E.A.; Evans III, R.B. *Theory. J. Chem. Ed.,* **1969**, *46(6),* 358-364.
24. Klinkenberg, L.J. *The Permeability of Porous Media to Liquids and Gases in Drilling and Production Practice;* American Petroleum Institute: New York, NY, 1941.
25. Stronestrom, D.A.; Rubin J. *Water Resour. Res.* **1989**, *36(4),* 1947-1958.
26. Reinecke, S.A.; Sleep, B.E. *Water Resour. Res.* **2002**, *38(12),* 16-1 - 16-15.
27. Penman, H.L. *Journal of Agricultural Science* **1940**, *30,* 437-462.
28. Millington, R.J.; Quirk, J.P. *Transactions of the Faraday Society* **1961**, *57,* 1200-1207.

29. Currie, J. A. *Monogr. Soc. Chem. Ind.* **1970**, *37*, 152-171.
30. Lai, S.; Tiedje, J.M.; Erickson, A.E. *Soil Sci. Soc. Amer. J.* **1976**, *40*, 3-6.
31. Farmer, W. J.; Yang M. S.; Letey J.; Spencer W. F. *Soil Sci. Soc. Amer. J.* **1980**, *44*, 676-680.
32. Thomson, K. A. M.Sc. thesis, University of Arizona, AZ, 1985.
33. Wakoh, H.; Hirano, T. *J. Loss Prevention in the Process Industries.* **1991**, *4* 260-264.
34. Fuentes, H.R.; Polzer, W.L.; Connor, J.A. *J. Environ, Qual.* **1991**, *20*, 45-51.
35. Johnson, R. L.; Perrott, M. *Journal of Contaminant Hydrology* **1991**, *8*, 317-344.
36. Hutter, G. M.; Brenniman, G.R.; Anderson R.J. *Water Environmental Research* **1992**, *64(1)*, 69-77.
37. Moldrup, P.; Olesen, T.; Schjonning, P.; Yamaguchi, T.; Rolston, D.E. *Soil Sci. Soc. Am. J.* **2000**, I, 94-100.
38. Karimi, A. A.; Farmer, W.J.; Cliath, M.M. *J. Environ. Qual.* **1987**, *16(1)*, 38-43.
39. Ehlers, W.; Letey J.; Spencer, W.F.; Farmer, W.J. *Soil Sci. Soc. Amer. Proc.* **1969**, *33*, 501-504.
40. Ehlers, W.; Farmer, W.J.; Spencer, W.F.; Letey J. *Soil Sci. Soc. Amer. Proc.* **1969**, *33*, 505-508.
41. Scanlon, B.R.; Nicot, J.P.; Massmann, J.W. In *Soil Physics Companion;* Warrick, A.W., Ed.; CRC Press: Boca Raton, FL, 2002; 297-34 1.
42. Massmann, J.W. *J. Environ. Eng.* **1989**, *115*, 129-149.
43. Baehr, A.L.; Hult, M.F. *Water Resor. Res.* **1991**, *27*, 2605-2617.
44. Weeks, E.P. Field determination of vertical permeability to air in the unsaturated zone. *USGS Prof. Paper* **1978**, *41*, USGS.
45. Nilson, R.H.; Peterson, E.W.; Lie, K.H.; Burkard, N.R.; Hearst, J.R *J. Geophys. Res.* **1991**, *96(B13)*, 21933-21948.
46. Edwards, K.B. *J. Environ. Eng* **1994**, *120(2)*, 329-346.
47. Werner, D.; Hohener, P. *Environ. Sci. Technol.* **2003**, *37*, 2502-2510.
48. Raney, W.A. *Soil Sci. Soc. Am. Proc.* **1949**, *14*, 61-65.
49. McIntyre, D.S.; Philips, J.R. *Aust. J. Soil Res.* **1964**, *2*, 133-145.
50. Rolston, D.E. Gas diffusivity. In *Methods of Soil Analysis;* Klute, A. Ed.; Am. Soc. Agron.: Madison, WI, 1986; pp 1089-1102.
51. Kreamer, D.K.; Weeks, E.P.; Thompson, G.M. *Water Resour. Res.* **1988**, *24*, 331-341.
52. Nicot, J.P. M.Sc. thesis, The University of Texas, Austin, TX, 1995.

Chapter 11

The Measurement and Use of Contaminant Flux for Performance Assessment of DNAPL Remediation

Michael C. Brooks and A. Lynn Wood

National Risk Management Research Laboratory, Office of Research and Development, U.S. Environmental Protection Agency, Ada, OK 74820

A review is presented of both mass flux as a DNAPL remedial performance metric and reduction in mass flux as a remedial performance objective at one or more control planes down gradient of DNAPL source areas. The use of mass flux to assess remedial performance has been proposed due primarily to the limitation of current remedial techniques to completely eliminate DNAPL from source zone areas. Four techniques are discussed to measure mass flux: the traditional method, integral pumping tests, passive flux meters, and horizontal flow treatment wells. Several different modeling approaches have been used to investigate changes in mass flux due to remedial activities. In general, these models indicate that three aspects of any given DNAPL-contaminated site are important to assessing the benefits of DNAPL remediation in terms of the flux response: the heterogeneity of the flow field, the heterogeneity of the DNAPL distribution, and the correlation between the two.

Nonaqueous phase liquids (NAPLs) have been a recognized threat to groundwater resources for roughly the past twenty years. Due to their physiochemical properties, they can persist in the environment for long periods of time and can generate expansive aqueous contaminant plumes. They are classified as either light NAPLs (i.e., LNAPLs) when they are less dense than water or dense NAPLs (i.e., DNAPLs) when they are more dense than water. The distribution of NAPLs in porous media can be a complex pattern depending on a number of factors, such as the spill history, the NAPL physical and chemical properties, and the hydrogeology and biogeochemistry of the environment in the spill vicinity. In general, the distribution of NAPLs can be described as a collection of fingers and lenses, the former referring to locations of residually trapped NAPL resulting from vertical migration pathways and the latter referring to locations at the end of the fingers where NAPL saturations are equal to or higher than residual saturation. Typically, LNAPLs are more easily managed than DNAPLs because they are less difficult to locate in the subsurface environment and because LNAPL constituents are often amenable to natural attenuation processes. In contrast, DNAPLs are more difficult to locate in the subsurface environment, are often recalcitrant to biodegradation processes, and, as a result, are the focus of much of the current NAPL-related research. Thus, this discussion will emphasize issues of particular importance for DNAPL source remediation.

The fundamental goal of DNAPL source treatment should be to reduce the risk posed by the DNAPL contamination, and Figure 1 provides a conceptual illustration of the relationship between DNAPL treatment and risk response. While quantitative risk analysis techniques are available and have been used in the management of contaminated sites, aqueous concentrations are more frequently used as a surrogate measure of risk and are the primary metric for regulatory decisions. Consequently, the principle motivation for the application of DNAPL treatment technologies is to achieve acceptable aqueous concentrations by either removing or destroying DNAPL mass in the source zone. Historically, groundwater treatment consisted of pumping contaminated groundwater from one or more wells and treating the effluent by a variety of water treatment technologies (i.e., pump-and-treat). Pump-and-treat approaches can be effective in plume management and are still widely used for that purpose. However, limitations of pump-and-treat systems to remove NAPL contamination (due, for example, to low NAPL aqueous solubility and heterogeneity-induced flow bypassing) which reduces contaminant aqueous concentration, have been recognized for over a decade (*1, 2, 3*).

As a result, a number of more aggressive remedial techniques were developed and tested for NAPL treatment. As an illustration, Figure 2 summarizes selected results from a number of technology demonstrations conducted under controlled conditions in isolated test cells at both Hill Air Force Base (AFB) and Dover AFB. The technologies shown in Figure 2 are cosolvent enhanced dissolution, (Cosolvent Dis.), cosolvent mobilization (Cosolvent

Mob.), surfactant enhanced dissolution (Surfactant Dis.), complex sugar enhanced dissolution (Corn. Sugar Dis.), and air sparging/soil vapor extraction (AS/SVE). While it was clear that these technologies could rapidly remove NAPL mass compared to pump-and-treat methods, they were unable to completely eliminate NAPL contaminant mass. Even in those cases where roughly 90% or more of the mass was removed, groundwater quality was still unacceptable compared to maximum contaminant levels within the test cells. Others have likewise noted the limitation of aggressive remedial technologies to completely remove DNAPLs (4, 5, 6) and have noted the inability to meet maximum contaminant level criteria, particularly in source areas, by partially removing mass from DNAPL source zones (7, 8).

Given the limitation of remedial technologies to meet acceptable water quality standards, particularly with respect to DNAPLs, the benefits of conducting costly source-zone remedial measures are unclear (5, 9). A number of source-treatment benefits have been suggested, and one such benefit is a reduction in contaminant flux from the DNAPL source area (6, 8, 10). If the flux that emanates from the source area is reduced by source treatment, conceptually, the contaminant plume may respond in a manner that reduces the contaminant risk, such as a reduction in total plume mass or a reduction in plume spatial extent. The flux response (Figure 1) to source treatment is critical to understanding the benefits of source treatment in terms of plume response. If the DNAPL mass in the source zone is reduced, will a reduction in contaminant mass flux occur? If so, will a beneficial plume response result, such as a reduction in the spatial extent of the plume or a reduction in total contaminant mass within the plume (i.e., a response that yields a reduction in the risk associated with the contaminant system)?

The use of mass flux or loading as a tool in groundwater characterization efforts has received increasing attention over the past decade. For example, Semprini et al. (11), Borden et al. (12), Kao and Wang (13), and Bockelmann et al. (14, 15) reported the use of mass flux measurements at multiple cross sectional transects along contaminant plumes to assess biodegradation activity. Einarson and Mackay (16) discuss the use of mass loading measurements to assess and prioritize contaminant impacts on down-gradient supply wells. Rao et al. (8), Soga et al. (6, 10), EPA (5), Stroo et al. (4), ITRC (17), and NRC (9) discuss the use of mass flux to assess benefits of DNAPL remediation. The purpose of this chapter is to provide a review of the mass flux work in the context of the conceptual relationships shown in Figure 1. Moreover, noting the importance of clearly defined remedial objectives and associated metrics (9), we adopt the view that contaminant flux reduction is a functional objective to the absolute objective of risk reduction. Modeling studies are reviewed to assess relationships between mass removal and mass flux. This is preceded by a discussion of field-scale contaminant flux measurements required to assess flux reduction.

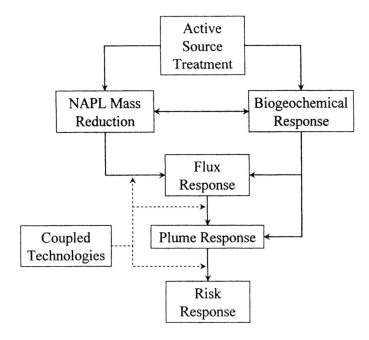

Figure 1. Conceptual illustration of source treatment and the associated system response.

Contaminant Flux as a Performance Metric

Metric Definition

Fundamentally, flux is defined as mass per area per time $[ML^{-2}T^{-1}]$:

$$\mathbf{J}(x,y,z) = c(x,y,z)\mathbf{q}_0(x,y,z) \tag{1}$$

where $\mathbf{J}$ is the flux vector $[ML^{-2}T^{-1}]$; c is the contaminant concentration $[ML^{-3}]$; and $\mathbf{q}_0$ is the Darcy velocity or specific discharge vector $[LT^{-1}]$ crossing a control plane oriented perpendicularly to the prevailing groundwater flow direction. Integration of the flux across a control plane yields mass loading (or discharge):

$$\dot{m} = \int_A \mathbf{J}\, dA \tag{2}$$

where $\dot{m}$ is mass loading $[MT^{-1}]$ and $\mathbf{A}$ is the control plane area $[L^2]$. The Darcy velocity is equal to:

$$q_0 = -ki \tag{3}$$

where k is the hydraulic conductivity $[LT^{-1}]$ and i is the hydraulic gradient [-].

One or more of several techniques can be used to measure contaminant flux in the field (*17, 18, 19*), and the available techniques to measure flux are classified here as the traditional approach, integral pumping tests, passive flux meters, and horizontal flow treatment wells. Most techniques invariably rely on independent estimates of concentration and flow velocity, although, currently, one technique does provide a direct estimate of contaminant flux (the passive flux meter). For conditions of heterogeneous hydraulic conductivity and contaminant concentrations, the scale-dependency of field measurements should be noted when comparing results obtained with different techniques.

Measurement Techniques

Traditional Approach

Mass flux can be estimated using independent estimates of concentration, hydraulic conductivity, and hydraulic gradients (see eq 1), all of which are typically measured during most site characterization efforts. The former is most often based on aqueous samples collected from monitoring wells. The latter two, used to estimate the Darcy velocity, are often estimated through independent field methods (i.e., slug or pump tests to estimate hydraulic conductivity and head measurements to estimate the hydraulic gradient). Alternatively, tracers can be used to estimate the Darcy velocity in one of two ways. The Darcy velocity can be estimated based on the displacement of the tracer as a function of time, or it can be estimated from borehole dilution tests. Borehole dilution tests are based on the dilution of a tracer as a function of time in an isolated portion of the well bore (20, 21).

Groundwater samples obtained from relatively long screen intervals are themselves flux-averaged concentration measurements and may differ from flux measurements based on samples collected over different spatial scales. As an alternative, multilevel sampling devices can be used to provide higher resolution spatial sampling. For example, Guilbeault et al. (22) present results from high resolution groundwater sampling using a direct push technique. In this study, hydraulic conductivity estimates were based on falling head permeameter tests using repacked core samples and from application of the Hazen formula to grain size analysis results. Hydraulic gradients were based on groundwater head measurements. Their results showed that for the three sites investigated, the spatial distribution of the contaminant mass discharge passing a control plane was highly heterogeneous, with 80% of the mass discharge located in 10% or less of the control plane area. Furthermore, concentrations varied between two to four orders of magnitude in 15 to 30 cm. These results suggest the level of detail that is required to accurately determine the contaminant flux across a control plane, and investigative efforts may yield results with high certainty only after a sufficient number of spatial samples have been collected and analyzed.

Integral Pumping Tests

Integral pumping tests (IPT) to measure contaminant flux were first presented by Schwarz et al. (23), and later discussed by Teutsch et al. (24) and Ptak et al. (25). Applications of the technique were presented by Bockelmann et al. (14, 15) and Bauer et al. (26). The technique has similarities to the traditional approach but consists of measuring contaminant concentration-time series from multiple pumping wells aligned perpendicular to the prevailing direction of groundwater flow. The use of concentration-time series information for site characterization efforts was apparently introduced by Keely (27), and the advantage of this approach is that the concentration-time series is measured in the pumping well effluent over much longer periods of time compared to the duration of traditional sampling techniques. Consequently, an IPT integrates information over a much larger volume, and therefore eliminates the need for a large number of sampling locations (15). Conceptually, under steady state conditions

$$\dot{m} = Qc \tag{4}$$

where Q is the steady state pumping rate $[L^3T^{-1}]$ and c is the steady state concentration. Equation 4 indicates that at steady state the mass capture by the pumping well is equal to the mass entering the capture zone of the well. However, such steady-state conditions may not be reached for tong periods of time. Under transient conditions, the concentration-time series information is used with an estimate of the Darcy velocity to calculate the contaminant mass flux at a control plane through the pumping wells.

Passive Flux Meters

Passive flux meters (28, 29, 30) consist of tracer-laden activated carbon packed inside a mesh material, which is placed in a well. The tracers are short-chain alcohols, such as methanol, ethanol, isopropanol, or isobutanol. The flux meter is left in the well for a duration, ranging for approximately 3 to 10 days depending on initial site estimates of Darcy velocity. While in place, the tracers are eluted from the sorbent into the flowing groundwater, and contaminants partition onto the sorbent. The flux meter is then removed and sub-sampled over discrete vertical intervals, and, based on the mass of remaining tracer and mass of accumulated contaminant, estimates of both the water flux (i.e., Darcy velocity) and the contaminant mass flux as a function of depth can be made. A similar approach was described elsewhere (31) but employed sparingly soluble salts as tracers and investigated several non-carbon absorbents.

By installing passive flux meters in a series of wells oriented perpendicularly to the prevailing groundwater flow direction, the mass flux across a control plane can be obtained. The advantage of this technique is that it provides a direct, independent measurement of both the Darcy velocity and the contaminant flux. Furthermore, it provides detailed information on the vertical distribution of the Darcy velocity and the contaminant flux. However, these meters sample a limited width of the aquifer. This width is a function of the convergence factor due to the increase in hydraulic conductivity of the well. Given the limited sampled width, relatively large areas in between welts are not sampled.

Horizontal Flow Treatment Wells

The use of horizontal flow treatment wells (HFTWs) for in-situ treatment has been described by McCarty et al. (*32*) and consists of two, dual-screened pumping wells, where the intake and discharge screens are reversed in the two pumping wells. The use of HFTWs to measure contaminant flux has been presented by Huang et al. (*33*) and Kim (*34*). The method is based on the measurement of interflow between the wells using a tracer, which provides a means to estimate the average hydraulic conductivity in the interrogated space. That information, combined with ambient hydraulic gradient data, then yields an estimate of the Darcy velocity. By measuring the contaminant concentration during tracer sampling, the contaminant flux can be estimated as well. A major advantage to this technique is that it integrates spatial information over a large area, like the IPT, but does not produce large volumes of contaminated water, which must be properly disposed.

Contaminant Flux Reduction as a Remedial Objective

Measurements of contaminant flux collected before and after remedial activities would provide an estimate of, presumably, flux reduction due to DNAPL remediation. Conceptually, a reduction in contaminant flux is expected to produce beneficial results due to the linkages of source zone, contaminant flux, plume response, and risk response (Figure 1). Sufficient field data to investigate changes in mass flux due to mass removal by source remediation are not currently available. In lieu of field data, a number of modeling studies are available to provide insight on the potential benefits of contaminant flux reductions.

Mass flux leaving DNAPL source areas is fundamentally a function of dissolution processes, and a great deal of work has been conducted on NAPL dissolution (see *6, 35, 36* for general discussions and reviews). The goal here is to review work more directly related to DNAPL mass-flux relationships. In the context of Figure 1, we first review an initial modeling study that explored the benefits of partial mass removal, then modeling studies that have investigated relationships between DNAPL mass reduction and flux response, and finally modeling studies that investigated plume response to changes in source mass and contaminant flux.

Benefits of Partial Mass Removal

White a number of numeric models have recently been used to investigate mass flux from source areas, they were preceded by an important study conducted by Sale and McWhorter (7) to investigate the benefits of partial DNAPL removal in terms of changes in aqueous concentration. They considered three-dimensional transport in a uniform flow field with rate-limited mass transfer between groundwater and an entrapped single-component DNAPL. In their approach, they were able to account for complex heterogeneous DNAPL distributions using superposition of analytical mass transfer solutions. They concluded that advective-dispersive transport is more limiting to mass transfer from the DNAPL subzones than the mass transfer rate coefficient, and, as a consequence, mass transfer from DNAPL subzones would be affected little by partial mass removal. It was also noted that rates of mass transfer were influenced by upstream and neighboring subzones, and, hence, the overall distribution of DNAPL in the entire source zone was important. Based on their results, Sale and McWhorter (7) suggested that the majority of the DNAPL would have to be removed to result in near-term, improved groundwater concentrations.

Two limitations of the method used by Sale and McWhorter (7) to reach their conclusions can be noted. Falta (37) demonstrated the importance of local-scale mass transfer by indicating that the rate of DNAPL dissolution decreased with time, compared to the constant rate of dissolution used by Sale and McWhorter (7). Another limitation of the analysis completed by Sale and McWhorter (7) is the assumption of a uniform flow field as noted by Rao and Jawitz (38). Under heterogeneous conditions, removing mass from areas of high flow may alter the mass flux, experimental evidence for which at least was provided by Nambi and Powers (39), who reported that variability in permeability in their two-dimensional laboratory models affected the extent of clean water flowing around the source zone and hence the flux-averaged concentration of contaminated water from the source zone.

DNAPL Mass Reduction/Flux Response Relationships

Subsequent modeling efforts have likewise demonstrated the importance of heterogeneity on mass removal-mass flux relationships. Soga et al. (6) investigated DNAPL mass removal-flux relationships using a 10,000-cell (each 0.5 m by 0.1 m), two-dimensional finite difference model of a 50 m long by 10 in deep source area (simulated release of 500 liters of perchloroethylene (PCE)). Multiple hydraulic conductivity field realizations were generated, each with the same statistical properties (in particular, a log-transformed hydraulic conductivity variance of 0.6). Results indicated that hydraulic conductivity and DNAPL distribution heterogeneity were more important to field-scale dissolution behavior compared to local mass transfer limitations, and multiple realizations with the same hydraulic conductivity statistical parameters produced significantly different flux-mass relationships. For example, the maximum mass discharge reported for a given hydraulic conductivity realization ranged from 40 mg/day to 100 mg/day. They also noted the importance of the morphological distribution of DNAPL compared to the actual DNAPL mass and noted that, within a given realization, significant changes in mass flux did not occur until there was a change in morphological distribution (i.e., mass associated with residual finger saturation was depleted leaving only pools).

Lemke et al. (40) investigated DNAPL infiltration and entrapment characteristics as a function of aquifer property correlations using detailed numeric simulations as well. That work was extended (41) to investigate surfactant remediation and mass flux relationships. Finite difference, two-dimensional modeling simulations were conducted on a heterogeneous (log-transformed hydraulic conductivity variance of 0.29) source area 7.9 m by 9.8 m in size, and the simulations included the release and redistribution of 96 liters of PCE. Multiple realizations of four ensembles based on alternative modeling formulations of porosity, permeability, and capillary pressure-saturation relationships were completed. They concluded that PCE effluent concentrations did not significantly change until the DNAPL architecture changed, specifically, until the predominant location of PCE mass was located in pools rather than residual fingers.

Parker and Park (42) conducted an investigation of heterogeneous (log-transformed hydraulic conductivity variance of 1) field-scale DNAPL (trichtoroethytene (TCE)) dissolution using high resolution (1,000,000 cells) finite-difference modeling simulations of a 10 m by 10 m by 10 m source-zone area. A total of 229 liters of DNAPL was allowed to infiltrate and spread under simulated conditions resulting in a complex distribution of fingers and lenses. A rate-limited mass transfer expression was used to model grid-scale DNAPL mass transfer, and the modeling results were used to estimate a field-scale mass transfer coefficient associated with the entire source area. Results indicated that apparent field-scale mass transfer coefficients were lower than laboratory-scale mass transfer coefficients reported in the literature, and this was attributed to heterogeneous distributions in flow velocity and DNAPL distribution.

Their results also indicated that the field-scale mass transfer coefficients varied in direct proportion to the mean groundwater velocity and that the functionality of the field-scale mass transfer coefficients to DNAPL mass could be approximated by a power-law expression. The exponent was less than unity for laterally extensive DNAPL lenses (i.e., pools), but greater than unity for randomly oriented residual DNAPL (i.e., fingers).

Lagrangian-based models have been used by Rao and Jawitz (38), Enfield et al. (43), Wood et al. (44), and Jawitz et al. (45) to investigate the benefits of NAPL remediation. The Lagrangian approach to contaminant flow modeling assumes the flow field can be represented as a collection of streamtubes, and the distribution of travel times within the stream tubes are representative of non-uniform flow under heterogeneous porous media conditions. Rao and Jawitz

(*38*) presented a simplified Lagrangian approach to investigate the relationship between NAPL mass removal and contaminant flux. They assumed a uniform NAPL distribution and a non-uniform flow field described by a log normal velocity distribution. The effects of partial remediation were investigated by assuming that, after a given period, a fraction of the streamtubes would be clean due to the remedial measures (i.e., those streamtubes with short travel times) and the remaining fraction would still contain contaminant mass (i.e., those streamtubes with large travel times). Assuming no variation in groundwater velocity (i.e., uniform flow), the contaminant flux would remain the same until alt stream tubes were clean (i.e., complete mass removal) as suggested by Sale and McWhorter (*7*). However, their analysis showed that, for small velocity standard deviations (log normal standard deviation of 0.1), the relationship between mass reduction and flux reduction was approximately a 1-to-1 relationship. Furthermore, as the velocity standard deviation increased (log normal standard deviation of 1.0), greater reductions in contaminant flux were achieved as a result of partially removing contaminant mass because of the influence of the high velocity stream tubes on the contaminant flux.

This approach was extended by Jawitz et al. (*45*) to include non-uniform NAPL distributions, rate-limited dissolution, and correlation between the velocity field and NAPL distributions. They presented closed-form expressions for mass and flux reduction in terms of parameters that could be measured in the field through NAPL partitioning tracer tests. The breakthrough curve of a non-reactive tracer is a measure of the stream-tube travel time distribution in a Lagrangian framework. Likewise, the breakthrough curve of a partitioning tracer can also be viewed as a distribution of travel times for a reactive tracer (*43, 45*), which is a function of both the groundwater velocity distribution and the NAPL distribution. Increases in the reactive travel time variance (resulting from either an increase in aquifer heterogeneity or NAPL distribution heterogeneity) required longer flushing times to remove NAPL mass but gave more favorable reductions in contaminant flux. A positive correlation between NAPL mass and groundwater velocity also gave more favorable results from the standpoint of greater reductions in contaminant flux for a given reduction in NAPL mass.

Enfield et al. (*43, 46*) and Wood et al. (*44*) described an inverse modeling approach using tracer data to develop site-specific Lagrangian-based models as well. Their approach used between 1 and 5 unique stream tube populations, which may or may not correspond directly to various hydrogeologic units. The travel time distribution for each population was represented as either a normal or log-normal distribution, and the NAPL distribution was assumed to be described by one of eight possible NAPL architectures, ranging from completely homogeneous to heterogeneously distributed (log-normal distribution) in only a fraction of the swept volume, in the case of multiple stream tube populations, each was assumed to have the same NAPL architecture. Calibration to site conditions was achieved by minimizing the difference between measured and modeled breakthrough curves for both conservative and NAPL partitioning tracers, as well as minimizing the difference between the first three normalized moments of the breakthrough curves. Enfield et al. (46) showed that under the conditions of uniformly distributed NAPL in a uniform flow field, little reduction in contaminant flux was achieved until most of the NAPL was removed. However, as the degree of flow heterogeneity increased, reductions in contaminant flux could be achieved by partially removing NAPL mass. Enfield et al. (43) demonstrated the technique using results from a partitioning tracer test conducted in isolated test cells from both Dover AFB and Hill AFB. The model was then extended (44) to predict remedial performance and the benefits of partial removal. Their results suggested that the benefits of partial NAPL removal in terms of flux response were a function of both NAPL architecture and hydraulic structure. The latter was noted to be a function of the pumping pattern employed during the tracer tests and the remedial tests.

Partitioning tracer test data can be used to calibrate models presented either by Jawitz et al. (45) or Enfield et al. (43) to site-specific conditions and to predict site-specific benefits of partial DNAPL mass removal in terms of flux reduction. However, it must be assumed that the flow fields under natural conditions, partitioning tracer test conditions, and remedial conditions are similar. Since the latter assumption may not be met for non-flushing remedial techniques, this approach is considered most applicable to flushing remedial technologies.

Analytical models have been used to express source concentration as a power-law function of the source mass (8, 42, 47, 48, 49). The power-law function presented by Falta et al. (48) is

$$\frac{C(t)}{C_0} = \left[\frac{M(t)}{M_0} \right]^{\Gamma} \tag{5}$$

where C_0 and M_0 are the initial source-zone concentration $[ML^{-3}]$ and mass $[M]$, respectively; $C(t)$ and $M(t)$ are the time-dependent source-zone concentration $[ML^{-3}]$ and mass $[M]$, respectively; and Γ is an exponential term representing the effects of the hydrogeochemical system (e.g., flow field heterogeneity and DNAPL saturation and morphology heterogeneity). In this approach, the source zone is treated as an inseparable unit, which is advantageous for looking at general system responses, especially considering the level of detail needed for more complex modeling work. The concentrations used in eq 5 can be viewed as the flux-averaged concentration crossing a control plane between the source area and dissolved plume. Figure 3 provides a conceptual illustration of the relationships between mass flux-averaged concentration and NAPL mass based on eq 5. For values of $\Gamma < 1$, the concentration declines less rapidly than NAPL

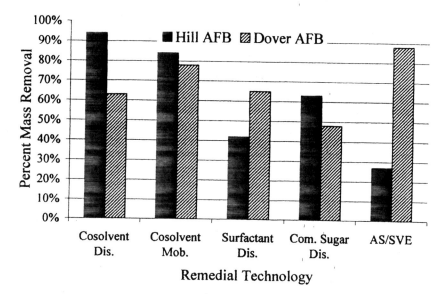

Figure 2. Selected results from aggressive NAPL remedial technology demonstrations. The Hill AFB contaminant was a complex LNAPL, and the percent mass removal was based on core data. The Dover AFB contaminant was a single component DNAPL, and the percent recovery was based on the mass balance. Even though relatively high mass removals were obtained, groundwater quality was not reduced to concentrations meeting maximum contaminant levels.

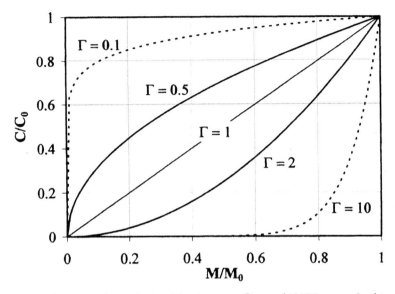

Figure 3. Power law relationships between flux and NAPL mass. In this representation, the origin represents perfect remediation (i.e., complete NAPL removal), and the upper right corner represents pre-remedial conditions.

mass, suggesting significant mass removal must be completed prior to reductions in concentration. For values of $\Gamma > 1$, the concentration declines more rapidly than NAPL mass, suggesting significant reductions in concentration may be possible with partial mass removal. Falta et al. (*48*) point out that in the latter case, while reductions in concentration are achieved with partial mass removal, the life of the source zone is considerably extended.

As expressed in eq 5, $C_0 \leq C_s$, which indicates that, while evidence suggests solubility limits are reached on local scales, on the whole, the source zone concentration wilt be less than the solubility limit. Both Parker and Park (*42*) and Zhu and Sykes (*47*) normalized the concentration using C_s and included terms to express the degree of non-equilibrium.

Jawitz et al. (*45*) suggested an alternative analytical expression based on their Lagrangian modeling work:

$$\Phi_{Flux} = \begin{cases} \Phi_{Mass}^{\Gamma}, & \sigma_\tau \leq 0.7 \\ \dfrac{\Phi_{Mass} + \beta\Phi_{Mass}}{1 + \beta\Phi_{Mass}}, & \sigma_\tau \geq 0.7 \end{cases} \tag{6}$$

where Φ_{Flux} and Φ_{Mass} are the reduction in flux and NAPL mass, respectively; σ_τ is the variance of the reactive travel times; and β is an empirical constant representing the effects of the hydrogeochemical system. The empirical coefficients were given as a function of σ_τ:

$$\Gamma = 1.31\left(\sigma_{\ln\tau}\right)^{1.22} \tag{7}$$

and

$$\beta = 1.03\left(\sigma_{\ln\tau}\right)^{4.50} \tag{8}$$

Mass Reduction/Flux Response/Plume Response Relationships

Falta et al. (*48, 49*) provided an analysis of general plume behavior as a function of flux response using idealized transport models. Equation 5 was used to relate flux response to DNAPL mass removal and was a boundary condition first to a transport model, considering $\Gamma = 1$, uniform flow (in the aquifer downgradient of the source zone), retardation, first-order degradation, and negligible dispersion (*48*). Likewise, Eq 5 was used to relate flux response to DNAPL mass removal and was a boundary condition to a second transport model, considering a range of Γ values, uniform flow (in the aquifer down gradient of the source zone), retardation, sequential first-order degradation and production, and three-dimensional dispersion (*49*). Results indicated that, when $\Gamma > 1$, partial DNAPL removal from the source zone reduces the concentration and mass within the plume, but the spatial extent of the plume is not significantly affected. The overall life of the source zone is also increased because the relative flux of mass leaving the source zone decreases with time. When $\Gamma < 1$, DNAPL source remediation does not significantly change the flux leaving the source zone, and consequently, the contaminant plume changes little. However, the overall source life is greatly reduced by mass removal. When $\Gamma = 1$, partial DNAPL mass removal produces a proportional linear reduction in plume mass and concentration.

Mass Flux and Remedial Strategies

Based on the preceding review of modeling work, three aspects of any given DNAPL-contaminated site are deemed of primary importance to assess the benefits of DNAPL remediation in terms of flux response: the heterogeneity of the flow field (σ_q, the spatial variance of the Darcy velocity), the heterogeneity of the DNAPL distribution (σ_S, the spatial variance of the DNAPL saturation and, for the purposes here, morphology), and the correlation between the two ($\rho_{q,S}$). While a continuum of flux responses is possible based on these parameters (*42, 45*), the following cases serve as a basis for relative comparison of sites with different characteristics:

- Case A: $\sigma_q = \sigma_S = 0$;
- Case B: $\sigma_q = 0$, $\sigma_S \gg 0$;
- Case C: $\sigma_q \gg 0$, $\sigma_S = 0$; and
- Case D: $\sigma_q \gg 0$, $\sigma_S \gg 0$ ($\rho_{q,S} < 0$ or $\rho_{q,S} > 0$).

Cases A and B are similar in that both consider a uniform flow field, but differ in that Case A represents a uniform DNAPL distribution and Case B represents a non-uniform DNAPL distribution. Case A is more representative in laboratory conditions, and appropriate remedial technologies are expected to perform well, resulting in significant mass removal and, possibly, acceptable aqueous concentrations. These conditions, however, are not representative of the more challenging field-scale DNAPL contaminant sites because rarely, if ever, are DNAPL and velocity uniformly distributed. Under Case B, the DNAPL is non-uniformly distributed but within a uniform flow field, which are the conditions for the model developed by Sale and McWhorter (*7*). This case is conceptually shown by the curve in Figure 3 with $\Gamma = 0.1$, and a large majority of the DNAPL would have to be removed to meet acceptable aqueous concentrations. Both the finite difference models and Lagrangian models have likewise supported this conclusion under these conditions.

Cases C and D represent non-uniform flow with uniform and non-uniform DNAPL spatial distributions, respectively. The former case was investigated by Rao and Jawitz (*38*), and reductions in flux as a function of mass removal were indicated based on the modeling results (i.e., $\Gamma > 1$ in Figure 3). In the latter case, both the Lagrangian and finite difference models indicated favorable reductions in flux as a function of mass removal. A positive correlation between velocity and DNAPL distribution resulted in a more favorable relationship between flux and mass (*45*).

Several conclusions based on conceptual modeling (*48, 49*) are important to overall DNAPL source zone remedial strategies. First, flux reductions relative to mass reductions (i.e., $\Gamma > 1$ in Figure 3) are considered favorable with respect to reducing the flux from the source zone. However, they are unfavorable from the standpoint of source longevity. Relative decreases in flux compared to mass reductions increase the source longevity because mass is removed more slowly from the source zone. Second, delays in source-zone remedial actions for new spills should be avoided. Source remediation has no impact on the mass of contaminant already in the plume and can only be viewed as a means to influence future plume response.

Finally, an important question to ask is how much flux reduction is sufficient to achieve the absolute goal of risk reduction as stated in the introduction. One approach which could be used to define the extent of required flux reduction is measurement of the site-specific natural attenuation capacity (*50*), which has been defined as the capacity to lower contaminant concentrations per unit length of flow in an aquifer and is a function of advective-dispersive and degradation processes. Once the contaminant flux has been reduced to a level at or below the site's capacity to degrade the contaminant, the plume is expected to respond in a manner consistent with risk reduction (*8*).

Conclusions

Predicting site-specific, mass removal-contaminant flux relationships would be most useful in the management of DNAPL-contaminated sites. At the present time, the means to make such predictions are limited and need validation through field tests. Under certain conditions, the Lagrangian approach based on partitioning tracer tests may be the most practical tool currently available for the characterization of site-specific mass flux-mass removal relationships. Certainly, more research is needed to efficiently predict site-specific mass removal-mass flux behavior. Once such predictions can be made with reasonable certainty, they wilt aid in estimating at which sites resources should be spent on aggressive DNAPL remedial measures. Sites at which significant flux reduction appears feasible relative to the natural attenuation capacity could be leading candidates for DNAPL source treatment.

References

1. EPA, Performance Evaluations of Pump-and-Treat Remediations, EPA/540/4-89/005, 1989.
2. Mackay, D.M.; Cherry, J.A. *Environ. Sci. Technol.* **1989**, *23,* 630-636.
3. Travis, C.C.; Doty, C.B. *Environ. Sci. Technol.* **1990**, *24,* 1464-1466.
4. Stroo, H.F.; Unger, M.; Ward, C.H.; Kavanaugh, M.C.; Vogel, C.; Leeson, A.; Marqusee, J.A.; Smith, B.P. *Environ. Sci. Technol.* **2003**, *37,* 224A-230A.
5. EPA, *The DNAPL Remediation Challenge: Is There a Case for Source Depletion,* EPA/600/R-03/143, 2003.
6. Soga, K.; Page, J.W.E.; Illangasekare, T.H. *J. Hazard. Mater.* **2004**, 110, 13-27.
7. Sale, T.C.; McWhorter, D.B. *Water Resour. Res.,* **2001**, *37,* 393-404.
8. Rao, P.S.C., Jawitz, J.W.; Enfield, C.G.; Falta, R.W.; Annable, M.D.; Wood, A.L. In *Groundwater Quality: Natural and Enhanced Restoration of Groundwater Pollution;* Thornton, S.F.; Oswald, S.E., Eds.; IAHS Press: Oxfordshire, UK, 2002; pp 571-578.
9. NRC, *Contaminants in the Subsurface: Source Zone Assessment and Remediation,* The National Academies Press: Washington, D.C., 2005, 372 pgs.
10. Soga, K.; Page, J.; Gowers, N. In *Environmental Geotechnics, Proceedings of the Fourth International Congress on Environmental Geotechnics, Rio de Janeiro, Brazil,* 2002; 2, pp 1069-1081.
11. Semprini, L.; Kampbell, D.H.; Wilson, J.T. *Water Resour. Res.* **1995**, *31,* 1051-1062.
12. Borden, R.C.; Daniel, R.A.; LeBrun IV, L.E.; Davis, C.W. *Water Resour. Res.* **1997**, *33,* 1105-1115.
13. Kao, C.M.; Wang, Y.S. *Environ. Geol.* **2001**, *40,* 622-630.
14. Bockelmann, A.; Ptak, T.; Teutsch, G. *J. Contam. Hydro.,* **2001**, *53,* 429-453.
15. Bockelmann, A.; Zamfirescu, D.; Ptak, T.; Grathwohl, P.; Teutsch, G. *J. Contam. Hydro.,* **2003**, *60,* 97-121.
16. Einerson, M.D.; Mackay, D.M. *Environ. Sci. Technol.,* **2001**, *35,* 66A-73A.
17. ITRC, *Strategies for monitoring the performance of DNAPL source zone remedies;* Interstate Technology & Regulatory Council: Washington, DC, 2004; 94 pgs.

18. American Petroleum Institute. *Groundwater Remediation Strategies Tool;* Publication Number 4730, API Publishing Services: Washington D.C., 2003; 80 pgs.

19. Nichols, E.; Roth, T. *LUSTLine Bulletin, New England Interstate Water Pollution Control Commission,* Bulletin 46, 2004; pp 6-9.

20. Halevy, E., Moser, H.; Zellhofer, O.; Zuber, A. In *Isotopes in Hydrology, Proceedings of the Symposium held by the international Atomic Energy Agency in Cooperation With the international Union of Geodesy and Geophysics,* Vienna, 1966; pp 531-563.

21. Palmer, C.D., *Journal of Hydrology* **1993**, *146*, 245-266.

22. Guilbeault, M.A.; Parker, B.L.; Cherry, J.A. *Ground Water,* **2005**, *43*, 70-86.

23. Schwarz, R.; Ptak, T.; Holder, T.; Teutsch, G. In *Groundwater Quality: Remediation and Protection;* Herbert, M.; Kovar, K.; Eds.; IAHS Press: Oxfordshire, UK, 1998; pp 68-71.

24. Teutsch, G.; Ptak, T.; Schwartz, R.; Holder, T. *Grundwasser,* **2000**, *5*, 170-175.

25. Ptak, T.; Schwarz, R.; Holder, T.; Teutsch, G. *Grundwasser,* **2000**, *5*, 176-183.

26. Bauer, S., Bayer-Raich, M.; Holder, T.; Kolesar, C.; Muller, D.; Ptak, T. *J. Contam. Hydro.* **2004**, *75*, 183-213.

27. Keely, J.F., *Ground Water Monit. Rem.* **1982**, *2*, 29-38.

28. Hatfield, K.; Annable, M.D.; Kuhn, S.; Rao, P.S.C.; Campbell, T. In *Groundwater Quality: Natural and Enhanced Restoration of Groundwater Pollution,* Thornton, S.F.; Oswald, S.E.; Eds.; IAHS Press: Oxfordshire, UK, 2002; pp 25-31.

29. Hatfield, K., Rao, P.S.C.; Annable, M.D.; Campbell, T. U.S. Patent 6,401,547 B1, 2002.

30. Hatfield, K.; Annable, M.; Cho, J.; Rao, P.S.C.; Klammler, H. *J. Contam. Hydro.* **2004**, *75*, 155-181.

31. De Jonge, H.; Rothenberg, G. *Environ. Sci. Technol.* **2005**, *39*, 274-282.

32. McCarty, P.; Goltz, M.N.; Hopkins, G.D.; Dolan, M.E.; Allen, J.P.; Kawakami, B.T.; Carrothers, T.J. *Environ. Sci. Technol.* **1998**, *32*, 88-100.

33. Huang, J.; Goltz, M.N.; Close, M.; Ping, L. In *Remediation of Chlorinated and Recalcitrant Compounds, 2004, Proceedings of the Fourth International Conference on Remediation of Chlorinated and Recalcitrant Compounds;* Gavaskar A.R.; Chen, A.S.C.: Eds.; Abstract 1D-08, Monterey, California, 2004.

34. Kim, S.J., Master's Thesis, Air Force Institute of Technology, Wright-Patterson Air Force Base, OH, 2004.

35. Feenstra, S.; Guiguer, N. In *Dense Chlorinated Solvents and other DNAPLs in Groundwater,* Pankow, J.F.; Cherry, J.A.; Eds.; Waterloo Press: Portland, OR, 1996; pp 203-232.

36. Khachikian, C.; Harmon, T.C. *Transport in Porous Media,* **2000,** *38,* 3-28.
37. Falta, R.W. *Water Resour. Res.,* **2003,** *39,* doi:10.1029/2003WR002351.
38. Rao, P.S.C., and J.W. Jawitz, *Water Resour. Res.,* **2003,** *39,* doi:10.1029/2001WR000599.
39. Nambi, I.M.; Powers, S.E. *J. Contam. Hydro.* **2000,** *44,* 161-184.
40. Lemke, L.D.; Abriola, L.M.; Goovaerts, P. *Water Resour. Res.,* **2004,** *40,* doi: 10.1029/2003WR001980.
41. Lemke, L.D., Abriola, L.M.; Lang, J.R. *Water Resour. Res.* **2004,** *40,* doi:10.1029/2004WR003061.
42. Parker, J.C.; Park, E. *Water Resour. Res.,* **2004,** *40,* doi:10.1029/2003WR002807.
43. Enfield, C.G.; Wood, A.L.; Espinoza, F.P.; Brooks, M.C.; Annable, M.; Rao, P.S.C. *J. Contam. Hydro.* **2005,** In Press.
44. Wood, A.L.; Enfield, C.G.; Espinoza, F.P.; Annable, M.; Brooks, M.C.; Rao, P.S.C.; Sabatini, D.; Knox, R. *J. Contam. Hydro.* **2005,** In Press.
45. Jawitz, J.W.; Fure, A.D.; Demy, G.G.; Berglund, S.; Rao, P.S.C. *Water Resour. Res.* **2005,** In Press.
46. Enfield, C.G.; Wood, A.L.; Brooks, M.C.; Annable, M. In *Groundwater Quality: Natural and Enhanced Restoration of Groundwater Pollution,* Thornton, S.F.; Oswald, S.E.; Eds., IAHS Publication Number 275, IAHS Press, Oxfordshire, UK, 2002, pp 11-16.
47. Zhu, J.; Sykes, J.F. *J. Contam. Hydro.* **2004,** *72,* 245-258.
48. Falta, R.W.; Rao, P.S.C.; Basu, N. *J. Contam. Hydro.* **2005,** *78,* 259-280.
49. Falta, R.W.; Rao, P.S.C.; Basu, N. *J. Contam. Hydro.* **2005,** *79,* 45-66.
50. Chapelle, F.H.; Bradley, P.M. *Biorem. J.* **1998,** *2,* 227-238.

Acknowledgement and Disclaimer

The authors thank the editors and the reviewer for their helpful comments during the review process. The work upon which this manuscript is based was supported by the U.S. Environmental Protection Agency through its Office of Research and Development with funding provided by the Strategic Environmental Research and Development Program (SERDP), a collaborative effort involving the U.S. Environmental Protection Agency (EPA), the U.S. Department of Energy (DOE), and the U.S. Department of Defense (DoD). This paper has been reviewed in accordance with the U.S. Environmental Protection Agency's peer and administrative review policies and approved for publication.

Indexes

Author Index

Subject Index

A

Acid mine drainage
  formation by biological and abiotic
    reactions, 106–107
  natural attenuation, 105–127
Acidithiobacillus ferrooxidans, 106,
  117
Acidobacterium capsulatum, 115
Acidophilic bacteria, 115–117
Acidovorax genus, 141
Advanced oxidation processes. *See*
  Fenton's reaction
Advective-dispersive local
  equilibrium solute transport model,
  34
Ametryn, sorption on organoclays,
  72–77
Amplified ribosomal DNA restriction
  analysis. *See* Restriction fragment
  length polymorphism
Anaeromyxobacter detection in
  subsurface by T-RFLP, 137
Analytical models and solutions,
  reactive transport, 157–161
  computer codes, 166–167
Aqueous solubility, toxaphene,
  methanol effect, 31–33
Arizona state Superfund site, field site
  for Fenton-dependent recovery
  GAC, 44–45*f*
Array technology, molecular method
  for environmental microorganism
  studies, 142–143
AT123D, analytical computer code,
  reactive transport, 166
Atrazine, sorption on organoclays, 72–
  77

Available metals concentration in soil,
  definitions, 16–17

B

Bacterial natural attenuation, acid
  mine drainage, 105–127
Bench-scale in Fenton-driven GAC
  regeneration experiments, 50–54
Beryllium availability
  distribution coefficient, 21–22
  mobile aqueous phase in soil, 19–
    20*f*
Beryllium concentration
  determinations, method comparison
  results, 17, 19*f*
Bifunctional organoclay, catalytic
  effect and mechanism in methyl
  parathion hydrolysis, 78–83
Binary diffusion coefficient,
  parameter for gas transport
  quantification, 190–192
  See also In-situ estimation gas
  transport parameters
Bioaugmentation in cleanup,
  petroleum-contaminated soils, 95–
  96
Bioavailability of contaminants, factor
  in bioremediation, 91
BIOCHLOR, analytical computer
  code, biodegradation and reactive
  transport, 166–167
Biodegradation and reactive transport
  analytical and numerical solutions,
  comparison, 168–171
  analytical modeling, 155–161
  computer codes for analytical